AUERBACH ON

OPTICAL
CHARACTER
RECOGNITION

AUERBACH ON series

Published:

Optical Character Recognition
Alphanumeric Displays

Coming:

Automatic Photocomposition
Microfilm Readers/Printers
Computer Output Microfilms
Retail Data Collection Systems
Industrial Data Collection Systems
Data Communications
Programming Aids
Facsimile
Digital Plotters and Graphic Displays
Small Business Computers
Key to Storage
Compilers
Software

AUERBACH ON

OPTICAL CHARACTER RECOGNITION

AUERBACH® publishers

princeton
philadelphia
new york
london

CONTENTS

PREFACE

This volume is one of a series of books covering significant developments in the information science industry. *AUERBACH On Optical Character Recognition* describes the elements making up optical character readers, the equipment and facilities required for their functional operation, differences between character and code readers, and developments in the technology, both past and present.

Eliminating the input bottleneck is one of the major concerns of computer users today. Optical character recognition (OCR), with its capability for high input speeds and reduction of input preparation time, is one of the most promising technologies for reducing input bottlenecks in the next two decades. In the past, OCR has developed rather slowly, owing to a number of problems that would be common to any device which attempts to deal with a relatively uncontrolled input medium encoded largely by personnel not specifically trained in the limitations of that device. In addition, the technology related to scanning and paper transport has been able to benefit from technological developments in other segments of the computer world only partially. This required a foundation of specialized research and development projects before an adequate technological base and supply of trained people could become available for the innovative and vigorously competitive atmosphere needed to bring down prices and raise the performance of character readers.

The number of new companies that have announced OCR products in 1969 and 1970 (in spite of the competition of key-to-tape devices) contributes to the conclusion that this initial period of slow development is at an end and that the 1970s will witness a great proliferation of numbers and types of ocr devices. The fact that some of these companies are small, recently formed companies organized for the purpose of marketing an ocr device reinforces this conclusion. This is an ideal point in time to take stock of the development of ocr in the past and explore the signs of the future in the activity of the present.

AUERBACH on Optical Character Recognition is an expansion of material derived from *AUERBACH Data Handling Reports* and *AUERBACH Standard EDP Reports*. These services are major units in the *AUERBACH Computer Technology Reports*, a looseleaf service recognized as the standard guide to edp throughout the world. They are prepared and edited by the publisher's staff of professional edp specialists.

Material in this volume was prepared by the staff of AUERBACH Info, Inc., and has been updated prior to publication; however, the field is changing so rapidly that the completeness of its contents cannot be guaranteed. For current information on data communications and significant companies offering products in this field, please refer to *AUERBACH Data Communications Reports*. Information can be obtained from AUERBACH Info, Inc., 121 North Broad Street, Philadelphia, Pa. 19107.

AUERBACH ON
OPTICAL CHARACTER RECOGNITION

1. INTRODUCTION TO TYPES OF EQUIPMENT AND MODES OF USE

Each day it becomes more evident that our large and complex computers do not operate in a void but in a social environment where they communicate with human beings as well as with other data processing equipment. Computer professionals have always known that the computer is only as good as the man who programs it, but they have not been always so keenly aware of the fact that computers are only as fast as the people who encode the data to be "digested." Through intensive research and development, the EDP industry has manifoldly increased the internal data processing capabilities of computers, but it has failed to develop input/output techniques and equipment to the same degree of efficiency. As a result, the human-to-machine communications gap, which has existed since the early days of automatic data processing, has further widened, limiting the potential throughput of today's high-speed systems.

This obvious speed differential between manual data preparation (such as keypunching and verifying) and computer processing has now reached a point where development of source-data automation and automated transcription devices has become essential. Recent developments in the peripheral equipment market, in particular the sudden expansion in manufacturers of magnetic-tape keyboard-encoding devices, have contributed to the increased awareness of the magnitude of the input/output "bottleneck" by highlighting the slow reading speed for cards and the problems of encoding with slow reading speed on buffered devices. Perhaps the most promising solution to the man/machine interface problem is the optical character reader (OCR), with its ability to digest information at a

1

rapid rate and, more importantly, to convert printed text or numeric data directly to a form suitable for computer processing, without intermediate conversion to coded media.

For the manager or the data processing professional who is just becoming acquainted with the bewildering numbers and kinds of input/output and data-encoding devices currently flooding the market, however, the character reader presents problems that are different from those he has previously encountered, and also introduces a somewhat unfamiliar technology. This book is addressed to the professional who has become interested in OCR but needs an overview of both the specialized technology and current equipment types in order to determine whether OCR is applicable to his problems.

Optical character readers are not new. The first OCR device was introduced in 1954 by Intelligent Machines Research Corporation (which later became part of Farrington Manufacturing Company). However, in spite of the high computer input speeds possible with those devices, acceptance of the optical character reader for data preparation has been slow. After 16 years, approximately 1000 readers were in use by 1970 in a worldwide population of more than 45,000 computer sites where keypunching still predominates.

Six factors have contributed to this slow rate of acceptance:

1. *High Cost to Users.* Optical character readers have been too costly to purchase for most installations and too costly to install. Extensive redesign of the data processing system, modification of input data preparation procedures, and retraining of personnel are all required.
2. *High Marketing Cost to Manufacturers.* Early manufacturers, primarily independents, lacked the resources to make the necessary marketing effort to penetrate the market. Users required not only indoctrination in the use of OCR, but also considerable technical support from the manufacturers.
3. *Failure of Manufacturers to Understand the Market.* Since early manufacturers did not recognize the large markets for OCR, they failed to design machines suitable for the markets. Instead, early design was geared to high-volume, special applications.
4. *User Fear.* Data processing managers, already pressed with burgeoning responsibilities and tight time schedules, hesitated to extend themselves still further to a new kind of system for which the probability of success appeared to be low.
5. *Technical Limitations.* Until recently (1968), manufacturers could not provide OCR machines with sufficient hand-print recognition capability. This capability is necessary so that the source-data re-

cording market can be penetrated significantly, since over 50 percent of source documents are hand-generated.

6. *Competing Technologies.* Key-to-tape or disk systems are in direct competition with OCR, and some users have converted to these systems in preference to OCR.

Further deterrents to acceptance of OCR have been the problems of those using OCR in the past. Good operation of an OCR system requires good control of the forms, the equipment, and the operators. Users have had difficulty obtaining from manufacturers the forms needed to satisfy the operating specifications of the OCR machines. Even when high quality paper was used and the forms were well designed, serious reading problems occurred when the forms were carelessly handled or the data were not carefully printed on them. As a result, users have had to work hard to achieve the control required for good operation.

Control-of-forms, quality equipment, and operator carelessness are only the beginning of the problems that affect successful recognition systems. Optical readers work on the principle of recognizing the difference in contrast between the character and the background on which it is printed. Many current character readers are severely limited by the type fonts they can read and, in some cases, by the size of the character set or sets they can handle. A character set is a group of related graphical symbols; e.g., the English alphabet is a 26-element character set, A through Z. Type font refers to the style and shape of the elements within the set. Some optical readers do not require special fonts and are capable (with suitable adjustments) of reading most fonts and even all fonts. So far, however, this capability is too expensive for widespread commercial use. The least expensive units are restricted to one font, which is usually especially designed for low error rates and is often limited to numerics plus a few special symbols.

MODES OF USE

Despite limitations in forms, font styles, and character sets that can be handled, OCR devices are being used effectively in many applications. The credit card, for example, lends itself well to such applications, and therefore many oil companies have converted to OCR systems. Utility billing agencies also rely to a great degree on the use of OCR equipment. In addition, some retail merchandising firms now use OCR techniques for charge account and billing applications. The United States government is using OCR equipment to process numerous types of documents, including employers' quarterly tax returns. The secret of the success of OCR in many cases is that, of all the general-purpose approaches to high-volume data input preparation, OCR is the only one capable of eliminating a specialized human operator for an intervening data transcription step. In this mode of operation—the DIRECT READ mode—the highest degree of success is obtained when documents come from sources that have imposed one or more controls on form, format, and type font. They include journal tape outputs of adding machines and cash registers, credit card transaction slips, airline tickets, carefully hand-printed meter-reading forms, etc. As these examples illustrate, the documents are of such diverse size, shape, and format (and occur in such high volume) that the paper-handling capability required of the OCR machines becomes a major cost element of the equipment for the DIRECT READ mode.

The likelihood of OCR being suitable for input data conversion in the DIRECT READ mode hinges on the degree of control over the source documents, whether they are originated in a central facility, in a field office, or by the public.

There are three levels of control over document preparation:

1. *Controlled documents* are generated in a central facility where the operators are well versed in the rules for document generation and are closely supervised.
2. *Semicontrolled documents* are generated in dispersed locations by operators familiar with the rules for generation but are not directly supervised.
3. *Uncontrolled documents* are generated by people, including the public at large, who are not familiar with the requirements for OCR input documents.

Fortunately, the majority of source documents are prepared in a central facility or dispersed company offices where some potential for control exists.

However, the DIRECT READ mode also imposes equipment design requirements that can result in excessive costs for OCR machines. These costs can be reduced if tight restrictions are imposed on the form and quality of the source documents. Such restrictions are not always achievable, so a second mode of operation has been developed to accommodate all types of documents. In this mode, quality control is achieved by completely retyping the input documents in order to reduce the requirements for complex OCR technology. In the RETYPE mode, source documents that do not measure up to machine-readable standards are completely retyped on an OCR input document by relatively low-cost labor (typists) to standards that are completely compatible with a given OCR machine. This machine can be relatively inexpensive because inherently it need handle only one size of form and be capable of reading only one stylized type font.

The RETYPE mode is a one-to-one alternative to the keypunching approach to data conversion. The three advantages of this mode are (1) that it does not require the highly trained labor for retyping and proofreading that is needed for keypunching; (2) the reading rate of ocr is higher than card readers; and (3) better control is achievable over system operation than with the DIRECT READ mode. With respect to the labor requirement, typing is faster than keypunching and this skill can be used generally around an office for many other tasks besides computer input data preparation. Thus, low-volume installations of ocr machines need not carry a high, fixed labor cost solely to support the machine. In fact, under peak loads, one can draw upon ordinary typing pools for ocr retyping.

With respect to operational factors, the documents to be read can be closely controlled by retyping. This procedure results in better performance of the ocr machine, i.e., a higher document acceptance rate and a lower rate of wrongly identified characters. It also provides design simplicity in the ocr machine, since only one type font must be recognized.

Both the quality of the input document and the data on it are important for proper operation of an ocr machine in both modes. Excellent reading performance can be achieved when:

1. The forms on which the data are printed conform rigidly to specifications.
2. The equipment printing the data on the input document is well maintained and carefully adjusted to specification.
3. The people creating the input documents are closely supervised to assure that they conform rigidly to the required procedures.

Furthermore, the greater the extent to which these conditions can be fulfilled, the simpler is the machine needed and the higher is the reading performance that can be achieved.

The capability to operate in the DIRECT READ mode is still the outstanding characteristic of the optical character reader. An ocr device designed for operation in the RETYPE mode must compete with several other types of readers, including code readers and magnetic-ink character readers, as well as new forms of high-performance buffered keypunches and keypunch replacements such as key-to-tape and key-to-disk encoders. Before exploring the distinctive features of the optical character reader, it is necessary to distinguish it from other classes of optical- and character-reading equipment.

DIFFERENCES BETWEEN CHARACTER AND "CODE" READERS

The main distinction between a character reader and a code reader is that a code is recognized by the relative position of the printed input on the paper, whereas a character is identified by its shape. The distinction obtains only in the larger sense, of course, since shape is determined by the relative position of a number of associated points, but it serves to separate the basically more complex problems of OCR recognition from those of a code reader.

Mark sense and bar code are the two basic types of optically sensed codes now on the market. Others include the binary "one" code used by the Cummins-Chicago Corporation Scanak, and the GE COC-5 characters (which are in an indeterminate state between character and code). New ideas include the printing of a code above and below a typewritten character in an easily recognized pattern. The Datatype Company has announced an inexpensive reader that reads such a code imprinted by a special Selectric typewriter ball, and Potter Instrument Company, Inc., has introduced a magnetic version. There are also remote terminals employing some of these codes. At present, code readers tend to be low-cost and low- and medium-speed devices, although there is no reason why high-speed devices could not be manufactured.

Mark-sense equipment senses the physical position or location of marks on a document, correlating the mark position to a previously defined equivalent character. Primary applications for mark-sense equipment are the grading of tests, coding of "exceptions" (payment of amounts other than those billed, for example), and in-the-field data entry for turnaround documents. As true optical character recognition techniques become more refined and less expensive, it is likely that mark-sense capabilities will gradually be replaced by handwriting capabilities. Most optical character readers provide mark-sense features in addition to character recognition.

Bar-code readers utilize thick line or bar representations of characters, a technique that limits their usefulness in human-oriented systems, since the bars cannot easily be read by human beings. In addition, only numeric characters can be represented in most bar codes. Bar-code readers are used by some oil companies, whose credit card imprinters print the bar codes onto the invoices at the point of sale. Recognition Equipment, Inc., also uses a bar-code imprint during the reading cycle to sort output documents. Examples of a number of codes are illustrated in Figure 1-1.

MAGNETIC vs. NONMAGNETIC READERS

Magnetic Readers

The second basic type of automatic character reader is the magnetic reader. Magnetic-ink character readers (MICR), which are used almost exclusively within the banking industry, can handle only special type fonts printed in magnetic ink. The font most widely used in the United States, and adopted as a standard by the American Bankers Association, is font E-13B—a highly stylized font that can be used to represent only ten numeric

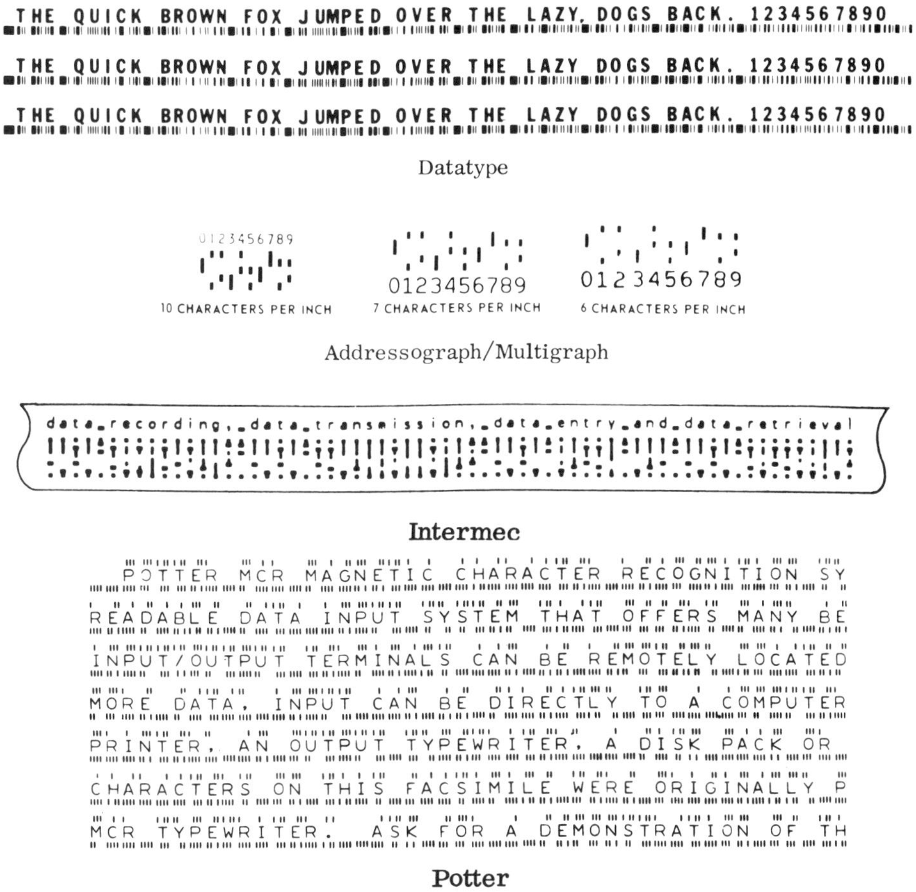

THE QUICK BROWN FOX JUMPED OVER THE LAZY DOGS BACK. 1234567890

THE QUICK BROWN FOX JUMPED OVER THE LAZY DOGS BACK. 1234567890

THE QUICK BROWN FOX JUMPED OVER THE LAZY DOGS BACK. 1234567890

Datatype

0123456789
10 CHARACTERS PER INCH

0123456789
7 CHARACTERS PER INCH

0123456789
6 CHARACTERS PER INCH

Addressograph/Multigraph

data recording, data transmission, data entry and data retrieval

Intermec

POTTER MCR MAGNETIC CHARACTER RECOGNITION SY
READABLE DATA INPUT SYSTEM THAT OFFERS MANY BE
INPUT/OUTPUT TERMINALS CAN BE REMOTELY LOCATED
MORE DATA. INPUT CAN BE DIRECTLY TO A COMPUTER
PRINTER, AN OUTPUT TYPEWRITER, A DISK PACK OR
CHARACTERS ON THIS FACSIMILE WERE ORIGINALLY P
MCR TYPEWRITER. ASK FOR A DEMONSTRATION OF TH

Potter

Fig. 1-1. Examples of printed codes.

digits and four special symbols (see Fig. 1-2). Another font, called the CMC-7 which was developed by Compagnie des Machines Bull-General Electric, is capable of representing all the characters in the alphabet as well as all the numeric symbols. However, the Bull font, although adopted as a standard by the European banking community, can be read by only some readers like the IBM 1259 and GE MRS-200 magnetic character readers.

Since magnetic readers detect only magnetic marks, nonmagnetic dirt or other marks will not cause reading errors. However, ink densities and character image are both critical, and the quality of printing on the docu-

Fig. 1-2. Magnetic reader font E-13B.

ments must therefore receive considerable attention to prevent character deterioration and extraneous ink spots. Quality control is already a problem within the banking industry and would be virtually impossible to enforce on a wider basis. Certain limitations in the use of MICR, however, have precluded its use in most commercial and industrial applications. The most important of these limitations is that magnetic ink must be used in all source document preparation equipment. Another is that the highly stylized E-13B type font used for MICR in the United States banking industry is unaesthetic and is limited to numeric sets. Despite these limitations, MICR equipment has proved reliable and useful in limited applications and will continue to be used for some time.

Classes of Nonmagnetic Optical Character Readers

Of the optical readers capable of reading printed characters or their coded equivalent, only the optical character reader is capable of directly reading a character by its outline or shape. These "true" OCR devices continue to develop along some of the lines laid down in the early 1960s. The OCR devices at that time included the following types of readers, classified by types of input documents and the transports needed to handle them:

1. *Document Readers.* These handle documents smaller than full-size 8.5 x 11-inch pages. The average size range is 3.0 x 3.5 to 4.0 x 8.0 inches. Document readers often read small character sets of numeric digits, plus a few special characters, in a special "turnaround filling" application that can utilize the DIRECT READ mode.
2. *Page Readers.* These handle a wide variety of document sizes including the standard 8.5 x 11 inches. Page readers usually read alphabetic, numeric, and special characters in the RETYPE mode, and may read more than one font in the DIRECT READ mode.
3. *Journal Tape Readers.* These read rolls of narrow paper variously called "listing tapes," "journal tapes," or "tally rolls." Journal tape readers usually read limited character sets (like those of document readers) at high speeds, often have multifont capabilities, and almost always read in the DIRECT READ mode.
4. *Reader Punch.* These read data printed on 80-column tab cards and punch the data into the same cards. Since no new model announcements have been made since 1966, and Farrington recently stopped producing one of the two available devices, this class of reader cannot be considered.

Between 1968 and 1970 two new classes of readers were added to existing categories as distinguished by input transports:

1. *Page-and-document readers* contain versatile page-reading transports capable of handling small documents at speeds sufficient to qualify them as document readers. They are often multifont readers capable of the DIRECT READ mode.
2. *Microfilm optical character recognition* (MOCR) devices have recently made their appearance in the form of the CompuScan, Inc., Model 370 and the Information International Corporation Graphics I, which are marketed as OCR devices that use off-line microfilming as an intermediary step between printed paper and the scanner. Since they are highly versatile omnifont readers, they are quite expensive. However, microfilm reading and handling capabilities need not be included in a high-cost reader. The idea is too new to predict its effect on the market, but it may lead to the missing link in a total microfilm input/output system as well as to a new way to handle paper.

Recent advances in microfilm technology and faster computer output microfilmers (COMS) support the possibility of a new movement in favor of microfilm as a business data-handling medium. Present MOCR readers probably do not handle COM output reliably because of the low resolution of COM characters. All readers mentioned above can also be subclassified into three basic categories according to type of configuration:

1. On-line systems that interface to a computer input/output channel and cannot encode an output medium independently of the computer.
2. Off-line systems that have an integral controller and output device. Many multifont readers are like off-line satellite computers that may optionally operate on-line or may optionally communicate, but which essentially have all control, recognition, and output logic inherent within the system.
3. Remote OCR (ROCR) terminals, which transmit characters (usually by a facsimile method) to a recording device at a remote location, where recognition occurs. At present, OCR remote terminals read only paper documents and cards, but MOCR terminals could conceivably work in a similar manner.

The use of a manual feed on Infonton, Inc., and Cognitronics Corporation readers adds another dimension to these classification schemes. Both manually fed devices are, naturally enough, among the cheapest devices on the market, since the transport is one of the most critical items in the price of a device. In the future there will probably be a considerable variety of manually fed or slow-moving transport systems to accommodate smaller users.

2. TRANSPORT AND RECOGNITION PROBLEMS: THE DIFFERENTIATION OF TYPES

The problems in handling differences in fonts are analogous to problems in handling a number of input codes. These are problems with which every computer user is familiar, and therefore he has little difficulty in grasping the significance of the great variety of input fonts and their relationship to the high price of optical readers. Anyone who has used a Xerox machine can also quickly anticipate the problems connected with a device that must move paper quickly and steadily. The newcomer to OCR invariably becomes puzzled, however, when confronted with the traditional distinction between page readers and document readers. He notices particularly that the lower limit of the range of paper sizes read by a "page" reader often dips well into the size range of paper sizes read by a "document" reader. At first glance the breakdown seems to be an arbitrary distinction between devices that read large pages and small pages, or large documents and small documents, if you prefer, since pages are often referred to as large documents.

The difference between the devices is not merely rhetorical, however, but is due to the nature of the scanner and transport, which have been designed particularly to meet the turnaround billing application. Before explaining the nature of these differences it is necessary to explore the three main parts that make up an OCR device.

OCR FUNCTIONAL UNITS

All existing commercial optical character readers consist of three basic functional units: an input transport, a scanner, and a recognition unit. A functional diagram of these three modules is presented in Figure 2-1.

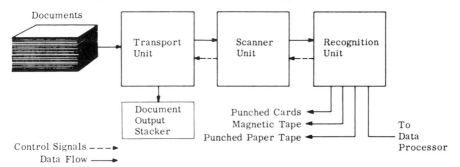

Fig. 2-1. Functional diagram of an optical character reader.

The function of the reader's *input transport* is to move the input to the reading station, position it properly, and then move it into an output stacker or, rather, into paper or card transport mechanisms. These mechanisms are classified into two basic types: One handles individual documents (paper sheets or cards) and the other handles continuous rolls (cash register or adding machine tapes). Microfilm transports differ slightly, depending on whether they handle sprocketed or unsprocketed microfilm. Most present-day readers employ the individual-document type of transport. It is estimated that ultimately the speed of all OCR systems will be governed by the speed of their input transports.

The function of a character reader *scanner*, like the function of a human eye, is to convert the optical images of the characters and symbols on a document into some analog or digital representation that can be analyzed by the recognition unit. The sensitivity, speed, and flexibility of the scanner are among the most significant factors in determining the quality of the whole system. Since there is a considerable range of capabilities, there are also various prices for the scanners.

The *recognition unit* is the brain of the character reader. This unit compares patterns from the scanner against reference patterns stored in the machine and either identifies the patterns as specific characters or rejects them as unidentifiable marks. Since the recognition unit is usually either a small special-purpose or a large general-purpose computer, this aspect of OCR technology has benefited enormously from the developments in computer technology. Lower prices, faster memories, and better software have all contributed to better and cheaper recognition units.

DOCUMENT TRANSPORTS

Input document transports in OCR journal tape readers designed to handle adding machine or cash register tapes consist of a tape well into which the paper roll is loaded, paper guides, and a paper drive control. Once the tape has been manually threaded, the paper is automatically moved past the read head, in somewhat like the movement of a film reel in a movie projector; a vacuum system is frequently used to keep the paper flat. The maximum length of paper roll that can be handled ranges from 100 feet to "any reasonable length."

In optical character readers designed to handle individual sheets or cards, the document-transport function is divided into two phases: feeding the documents from the input hopper and transporting the documents past the reading station. A common device for document feeding is the friction feeder, used in the IBM 1282 optical character reader. This device consists of a belt wound around capstans and partially resting on the document stack. Constant pressure is exerted against the belt by the document stack. As the belt moves across the top of the stack, it pushes the top documents into a separator station, where a combination of rollers and another belt separates the top document from all documents below it. Vacuum or suction feeders are sometimes used to lift documents off the input stack. Both the friction and vacuum devices, however, have problems in handling documents of thin paper and many occasionally feed more than one document at a time. A new type of feeder, designed by Rabinow Electronics (a subsidiary of Control Data Corporation), is said to eliminate the possibility of feeding two sheets at a time. This feeder uses a set of cone-shaped rollers that engage a corner of the topmost document and roll up the corner away from the pile so that it can be grasped by paper rollers that carry the document to the transport unit.

A popular method for transporting the document to the reading station is a vacuum-drive conveyor belt. Some character readers, such as the IBM 1428, use the conveyor belt to place the document on a rotating drum, which moves the document past the read head. The paper is held to the drum by means of a vacuum. The type of transport used at the reading station is affected by the scanning method. If the scanner can read only one character at a time, the paper-positioning movements must be more sensitive than those needed when the scanner can handle a line segment, a line, a paragraph, or page without requiring repositioning. If the scanner is fixed, the paper must move, but if the paper proceeds at a fixed rate, the scanner must move. In the case of a document containing one line of print in a predetermined position, the scanner can be fixed and the transport can move the characters in a steady flow past the scanner without requiring precise positioning of either unit. The disadvantage of all these mechanical techniques is that the paper cannot be moved and positioned as fast as it can be read.

The worst problem with mechanical paper handling is that paper transports tend to jam, even with good quality paper. Paper jams can be quite expensive, since the whole system must remain idle while the jam is cleared. Several approaches can overcome various problems of the transport unit. At lower speeds, it need not be so complex (and so costly). At higher speeds, moving air can be directed within the unit to reduce aerodynamic instability of the document in motion. CompuScan, Inc., has approached the problem by off-line conversion to negative microfilm images followed by higher-speed, jam-free OCR scanning techniques. This approach is based on the theory that time lost in the conversion process is more than regained by the higher speeds of microfilm transports and by transferring costs of downtime due to paper jams to a relatively inexpensive piece of equipment. Finally, since scanning means relative motion— i.e., moving the document or the viewing system—a high-resolution cathode-ray tube (CRT) can be used to scan an entire document electronically without requiring mechanical movement of the document except to bring it into scanning position, which is a relatively simple transport function.

The transport unit is an important cost component of an OCR system, typically involving 50 percent of the total system cost. Moreover, cost reductions through technological improvement tend to be limited by the mechanical nature of the problem.

SCANNER UNITS

Optical scanning methods are based on the differences in contrast between the characters and the background on which they appear. The function of the scanner is to sample either portions of a character or a complete character, to determine the relationships between light and dark areas. Dirt, the condition of the paper, paper reflectivity and ink-color smoothness, the age of the print, the accuracy of character registration, and the sharpness of character shape all have an effect on the scanner's ability to determine the relationship between the light and dark areas.

A large part of the cost of a flexible reader that can distinguish a number of fonts and deal with badly degraded documents is incurred by the sensitive and discriminating optical system. Readers that can handle one highly stylized font printed with OCR ribbons on special high-quality OCR paper can obtain a respectable throughput rate with a less expensive scanning system. Relatively inexpensive scanners can be used by MICR systems because recognition governed by the presence or absence of magnetic ink is not affected by dirt, reflectivity, color, and other factors that can affect light-versus-dark discrimination systems. Sensitivity, speed, and flexibility of the scanner are among the most significant factors in determining the quality of the whole system.

The common types of OCR scanners are mechanical disks, flying spots, parallel photocells, and vidicon scanners. All such units scan and convert a portion of a document to be read to electronic impulses that reflect the pattern of black and white, doing so either mechanically by movement of the document or of the mirrors of an optical lens system, or electronically by the movement of sensors over the document. The MICR scanner units also use varying techniques. Each of the common types of OCR scanners is described below, and the relative advantages and disadvantages of each are presented in Table 2-1.

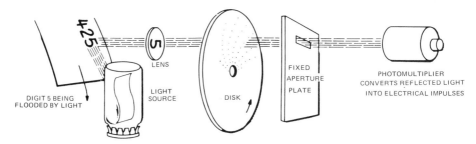

Fig. 2-2.   Mechanical disk scanner, Farrington readers.

Mechanical Disk Scanner

The mechanical disk scanner consists of a lens system, a rotating disk, a fixed aperture plate, and a photosensitive device like a photomultiplier tube or a solar cell, as shown in Figure 2-2. The characters to be read are flooded with light, which is reflected from the surface of the document onto a rotating disk via the lens system. The disk has apertures ranging from its center toward its periphery; as it rotates, these apertures pick up light samples. A fixed aperture plate regulates the amount of light and directs the light to a photosensitive device, which converts the light samples into signal pulses. By varying the voltage threshold, the photocell outputs can be adjusted for different background colors.

The mechanical disk scanner senses a character of data at a time. Movement between characters and lines is accomplished either by moving the document, as in the old NCR optical journal reader, or by repositioning the

Table 2-1. Comparison of Scanner Units.

TYPE	DESCRIPTION	ADVANTAGE	DISADVANTAGE
Mechanical	Consists of a lens system, a rotating disk, a fixed aperture plate, and a photomultiplier. The lens system directs light reflected from the characters through the rotating disk to form light samples detected by the photomultiplier, which converts them to electric signals.	Cost	Relatively slow speed
Flying-Spot Scanner	Consists of a cathode-ray tube (CRT), lens system, and phototube. The CRT scans the document with a beam of light, which the lens system directs to the phototube for conversion to electric signals.	Flexibility to scan selected portions of the documents	High cost
Parallel Photocells	Consists of a row of photocells. Characters are passed under the photocells, which simultaneously sample a number of elements of the character and convert them to electric signals. A more expensive variation uses a full matrix of photocells to increase speed.	High speed	Limited flexibility and high cost
Vidicon	Characters are projected onto a vidicon television camera tube, scanned by an electron beam, and converted to electric signals.	No requirement to move documents	Limits on number of characters stored on tube and limits in speed

lens system, as in the IBM 1428 alphameric optical reader. Consequently, this type of scanner is relatively slow in comparison to the other scanners mentioned.

Flying-Spot Scanner

The flying-spot scanner consists of a cathode-ray tube (CRT), a projection lens, a phototube, and a control unit. A beam of light is generated in the CRT and deflected across the tube in a scan pattern. The lens system projects this scanning light spot onto the document, from which it is reflected into a phototube. The phototube generates a voltage signal whose level is proportional in each instant to the amount of reflected light, thus indicating light and dark areas. The resulting signals are then either fed directly to the recognition unit in analog form or first transformed into digital form.

The flying-spot scanner offers more flexibility than the mechanical disk scanner, since its scanning pattern can be automatically adjusted by the control unit to permit the use of different scanning modes (i.e., scanning certain character fields or scanning specified portions of the document). Also, being completely electronic, the flying-spot scanner is faster than the mechanical disk scanner and is generally classified as a medium-speed device.

The introduction of high-resolution CRTs (2000 optical lines) has encouraged manufacturers to favor the development of a reader in which a complete document can be scanned without any document motion other than that required to position it under the read station. At present, however, such systems are quite expensive.

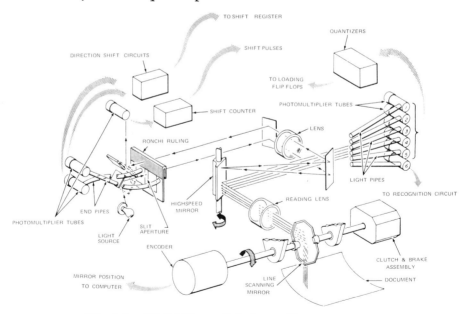

Fig. 2-3.   CDC 915 parallel photocell scanning system.

$\bar{2}$		2		$\bar{2}$
$\bar{2}$		1		$\bar{2}$
$\bar{2}$		2		$\bar{2}$
$\bar{2}$		1		$\bar{2}$
$\bar{3}$		1		$\bar{3}$
	1		1	
	1		1	
	1	1	1	
	1	1	1	
1		$\bar{2}$		1
1		$\bar{2}$		1

1	2	2	2	
1	2	2	2	
2				
2		$\bar{2}$		2
2		$\bar{2}$		2
2		$\bar{2}$		2
2		$\bar{3}$		2
2				
1	4	4	4	
1	4	4	4	
2				
2		$\bar{3}$		2
2		$\bar{2}$		2
2		$\bar{2}$		2
2		$\bar{2}$		2
1				
1	2	2	2	
1	2	2	2	

Fig. 2-4. Shift register contents for the letters A and B, showing the weighted areas in their respective recognition matrices superimposed on the "flip-flop" positions of the shift registers.

Parallel Photocells

The use of a vertical column of photocells speeds up scanning operations by simultaneously sampling a number of points that; when combined, add up to a complete vertical slice of the character. The electric signals generated by the photocells are then quantized into black, white, or gray levels, and these data are fed into a shift register where they are stored until data on the entire character have been accumulated. Owing to the simultaneous parallel sampling, this type of scanner can achieve higher speeds than the flying-spot scanner. The Control Data 915 utilizes a column of photocells, as illustrated in Figure 2-3; the data stored in the shift registers are represented in Figure 2-4.

One variation of this sampling method eliminates the need for shift registers by using a full "retina" or matrix of photocells to sample an entire character rather than just one vertical slice. Besides eliminating the shift register, this method also increases reading speed to approximately 2400 characters per second. Recognition Equipment, Inc., is one of the companies currently using a retina of photocells for sampling. This sampling technique has the present capability of achieving a higher speed than any of the previously mentioned techniques.

Figure 2-5 shows a degraded character and what the Recognition Equipment Electronic Retina® machine "sees." Unfortunately this method is quite expensive because as many as 800 photocells may be needed for alphanumeric character sets.

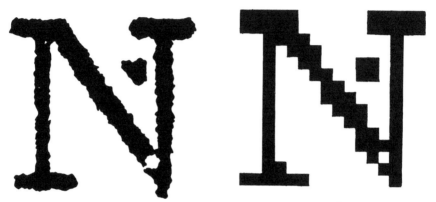

Fig. 2-5.   Electronic Retina® character sampling.

Vidicon Scanner

So far, this discussion has concerned scanning methods that read characters by reflecting light from the document to one or more photocells. A totally different method is to project the characters onto a vidicon television camera tube and scan the active surface with an electron beam. The resulting video signals are quantized to indicate black or white digitally.

A scanner that operates in this manner has been used by the UNIVAC Division of Sperry Rand Corporation. Since a group of characters can be stored on the tube (the NDP vidicon scanner can store 45 characters), there is no need for document movement during the scanning operation if the document contains a reasonably small number of characters. The advent of high-resolution vidicon tubes permits the character capacity to be eliminated on most documents.

Another advantage of the vidicon scanner is speed. Since it takes only 30 milliseconds for the beam to scan the entire tube, a full grouping of stored characters can be read in that time. At present, owing to the limited number of characters that can be stored on the tube, the scanner has only medium speed; i.e., it scans about 500 characters per second. However, once this limitation is removed, vidicon scanners should be as fast or faster than the flying-spot type.

Magnetic Scanner Units

Since the banking field represents the major application area for magnetic character readers, all magnetic readers produced in the United States have scanning units designed to handle the E-13B font shown in Figure 1-2.

Most scanning units convert the magnetic characters into an analog voltage waveform for subsequent identification. When a character moves past the read head, it generates an electric signal. This signal has a unique waveform, which the recognition unit matches against reference waveforms. The companies presently using this technique are Burroughs, General Electric, and National Cash Register.

The digital scanning technique used by IBM is exemplified by the IBM 1419 magnetic character reader. In this machine, each character is scanned by 30 magnetic heads, stacked vertically and interconnected to give ten outputs. The outputs are transmitted to a 70-bit shift register in the recognition unit, where they are matched against stored reference patterns.

RECOGNITION UNITS

Recognition units probably represent the area of greatest technical development in the character-reader field. Because of the rapid progress in this area, this discussion is restricted to the more common types of recognition techniques available commercially. Note that the recognition logic need not be limited to one of these techniques, particularly when recognition is largely under software control. The Control Data 935 document reader uses physical curve tracing on a numeric hand-print character and a raster scan on an alphabetic hand-print character. A comparison of the basic advantages and disadvantages of the main types is presented in Table 2-2.

Matrix Matching

This technique, one of the more widely used, stores the scanner signals in a digital register that is connected to a series of resistor matrices, each of which represents a single reference character. The other end of each matrix is connected to a second digital register, whose voltage outputs are representative of what would be obtained if the reference character were present. Recognition is based upon the resultant output voltage obtained from each matrix.

Table 2-2. Comparison of Recognition Units.

TYPE	DESCRIPTION	ADVANTAGE	DISADVANTAGE
Matrix Matching	Consists of digital registers and resistor matrices representing each recognizable character. The electric output of the scanner is stored in the digital register and compared electrically to each resistor matrix. Recognition is based on the relative comparison among the matrices. Can also be implemented by comparing the binary image of a character with images stored in memory.	Speed; flexibility to change character sets and to identify full alphanumeric character set	Cost
Stroke Analysis	Consists of a special-purpose computer. Recognition is based upon analysis of the stroke or line formation of each character in comparison with stored information for each recognizable character.	Ability to identify full alphanumeric character set	Low speed; unsuitable for hand printing
Curve Tracing	Each character's outline curvature is traced by either movement of the scanner beam or by logical decisions of the recognition unit. Recognition is based on such features as character splits, line junctures, line directions, straightness of lines, and line magnitude.	Suitable for hand-print numerics	Difficulty in reading certain alphanumeric characters

The advantages of the matrix-matching technique are that the resistor matrices can be modified so that character fonts can be changed easily and a full alphanumeric character set can be read. The technique also has the advantage of being quite fast (up to 2400 characters per second), since the matching is done by resistor matrices. The technique is similar in theory to the optical-matching technique described earlier, but it can handle misregistered characters much more effectively. The numerous machines using this technique are listed in the comparison chart (Tables 5-2 and 6-3).

Stroke Analysis

This technique, previously used by Farrington Manufacturing Company, is based on the stroke or line formation of each character. The characters are differentiated by the number and position of vertical and horizontal strokes. A special-purpose computer matches the formation of the unknown character against a character truth table, which indicates the stroke formation for each reference character.

Stroke analysis has the advantage of being able to handle a full alphanumeric character set, but the maximum speed obtainable by the Farrington character readers is about 600 characters per second, which is low compared with the 2400 characters per second obtained by machines using the matrix-matching technique. Also, the stroke-analysis method does not have the font flexibility of the matrix-matching technique because the wired recognition program in the special-purpose computer must be changed whenever there is a switch to a different character font.

Curve Tracing

The curve-tracing technique is a new concept. Since it can accommodate relatively wide variations in the shapes or sizes of the characters, this technique is useful for recognizing handwritten characters. By tracing the character's curvature, the recognition unit determines certain features that are used to identify the character. It records and evaluates such variables as character splits, line junctures, line directions, straightness of lines, and line magnitudes. The major disadvantage is difficulty in reading characters that have broken lines or holes within the black boundaries.

Tracing a character outline can be accomplished either by physically moving the scanner or by logical decisions of the recognition unit. Physical scanning involves following the black outline of the character by using a circular motion or step function to zigzag between its black and white boundaries. The recognition unit directs the scanner to move in specific directions based on what has already been detected.

Since physically following the character outline is both complex and time consuming, a better curve-tracing technique is to perform a raster scan uniformly across the character, transmitting contrast information to the recognition logic, which stores the data and determines changes in direction of the character with respect to the last scan. A "mental" picture of the character shape is thus developed within the logic unit. Character splits (such as the intersection of the horizontal and vertical lines of the letter "H") cause the logic to follow either line according to prescribed rules.

Analog Waveform Matching

Analog waveform matching is another recognition method that has been in use for some time, particularly in the magnetic-character readers used by the banking industry. It is based on the principle that each character passing under a read head produces a unique voltage waveform as a function of time; that is, the waveform of each character differs in either shape or length with respect to time. Characters are identified by matching their waveforms to reference waveforms. Machines using this technique have reading speeds of approximately 500 characters per second. The principal disadvantage of this system is that only a limited number of characters have unique waveforms; therefore this technique is found mainly in systems dealing with a limited character set.

Frequency Analysis

Frequency analysis is a digital recognition method developed for fonts consisting of closely spaced vertical lines. The widths of the gaps between the vertical lines of each character are measured by variations in magnetic flux. An unknown character is identified by comparing the sequence and number of its narrow and wide gaps with stored codes for each of the alphanumeric characters. An analog version is currently being used in the General Electric MRS 200 document reader.

The advantages of the frequency-analysis technique include the ability to accommodate a full character set, and increased reading speeds.

SPECIALIZED SCANNING AND TRANSPORT FOR TURNAROUND BILLING

The frequent use of a particular type of billing has led to development of a particular type of OCR device to fit that application, namely, the document reader. Briefly, in a turnaround billing a small computer-generated bill is mailed out to a customer, the customer returns both the bill and his payment, and the document is read into the computer again to record the payment. Since the account number, customer information, date, amount of bill, and other details are relatively small amounts of data, they can usually be printed in a single line. This means that the scanner can be fixed in a predetermined location and the small documents can be fed in at high speeds. Often a numeric character set and a single font are all that is necessary to handle the application.

A device made to handle a variety of papers that have been fully printed with data in various locations, on the other hand, cannot be ex-

Table 2-3. Differentiation Between Conventional Page and Document Readers.

CHARACTERISTIC	PAGE READER	DOCUMENT READER
Input Document Specifications		
Typical size	8.5x11, occasionally larger (11x14) or smaller.	Tab card size or smaller (bills, coupons).
Ideal printing format	As many lines as can be crammed on the scanning area; usually double-spaced to reduce errors.	One or two lines of type in predetermined locations, occasionally up to five lines per document.
Typical character set	Alphanumeric.	Numeric.
Typical application	Varies widely.	"Turnaround" billing.
Transport and Scanning Systems		
Scanner (flexibility)	Usually highly flexible and in delicate balance with transport.	Relatively inflexible; usually fixed in one or two predetermined locations.
Transport system requirements	Capable of steady movement and exact positioning of input document; responsive to control logic.	Capable of moving documents steadily and uniformly at high speeds without jamming or double feeding.
Speed		
Significant factor	Scanning speed.	Rate of movement of physical document.
Usual expression	Characters per second or lines per minute.	Documents per minute or documents per hour.
Data preparation	Large (expensive) multifont readers may read unprepared material in compatible fonts, but in many cases benefit from retyping; smaller, single (stylized) font readers require retyping of source document before reading.	Source document prepared by computer in stylized type font in rigid format; after being used to govern transaction (e.g., utility bill, meter reading), appropriate discrepancies are noted, using mark sense or hand print, and source document is read.

23

Table 2-4. Representative Types of Current Generation OCR Machines.

MACHINE	TYPE I	TYPE II	TYPE III	TYPE IV	TYPE V	TYPE VI
Type of document reader	Page	Page	Document	Document	Page-and-document	Journal tape
Mode of use	DIRECT READ	Retype	DIRECT READ	DIRECT READ	DIRECT READ, RETYPE	DIRECT READ
Font capability	Multi-font	Single stylized	Multi-font	Single stylized	Multi-font	Single stylized
Character set	Alpha-numeric	Alpha-numeric	Alpha-numeric	Numeric	Alpha-numeric, numeric	Numeric
Hand-print capability	Common	Occa-sionally	Common	Occa-sionally	Common	None
Document-handling speed	Low	Low	High	Moderate	High to low	High
Character-scanning rate	Moderate	Moderate	High	Moderate	High	High
Typical purchase prices	$400,000	$175,000	Varies widely; $100,000, $750,000	$120,000	Varies widely; $100,000, $750,000	Varies widely; $100,000, $175,000
Examples (delivered systems)	Scan Data 200; Compu-Scan 370	Control Data 915; IBM 1288	IBM 1287; REI In-put 2	Control Data 936-1; UNIVAC 2703	Control Data 955; Scan-Optics 20/20	NCR 420-2; Farring-ton 4040

pected to perform its function without either a scanner or a transport capable of precise positioning. Moreover, it is the speed of the searching and scanning and the capabilities for selective reading that will govern throughput more than the speed at which the paper moves. Obviously, the requirements of a transport that must interact very precisely with a scanner are different from those for a transport to be used for high-speed paper movement in which paper positioning is of only minimal concern. This differentiation is the basis of the distinction between the traditional page-and-document reader.

Needless to say, some businesses have both turnaround billing applications and other types of applications suitable to a page-reading device; in the past, this was solved by Recognition Equipment, Inc. (REI), by providing two input transports to a common recognition and output module, but the cost of the REI Electronic Retina® computing reader was prohibitive to all but the largest users. It was not until 1970 that Scan-Optics, Inc., and Control Data Corporation announced relatively fast, versatile transports capable of reading one- or two-line documents as well as complete pages at high speeds.

Table 2-3 presents the basic differences between traditional document and page readers, and also lists some of the common tendencies that have evolved in response to the basic differences in their applications. Table 2-4 summarizes the basic characteristics of the six representative types of devices offered on the market in 1970.

Before examining some of the specific page readers, document readers, and page and document readers offered by current manufacturers, the nature of two input problem areas common to all OCR devices needs to be discussed. These two problem areas are the form of the input data (i.e., fonts and character sets) and the printing of the medium itself (i.e., the effect of the printers, ink, paper, and forms design on OCR performance). These areas are discussed in the two subsequent chapters.

3. INPUT PROBLEMS I:
FONTS AND CHARACTER SETS

Character sets and type fonts take on a special significance when applied to the OCR field, since they govern the nature of the input and output media that can be accommodated. In general, OCR character sets and type fonts must

1. Represent all data to be read by OCR.
2. Retain OCR-readable characteristics despite some degradation of source documents.
3. Be legible and esthetically acceptable to human readers.
4. Be reproducible by present printing equipment, computer printers, adding machines, and typewriters.

CHARACTER SETS

In order to minimize the basic equipment cost and to decrease the error and reject rates of present OCR equipment, there has been an effort to minimize the size of the character set required for any given application. In general, restricted character sets fall into one of two categories:

1. *Numeric Sets.* Strictly numeric readers are commonly used to read imprinted documents or journal tapes produced by computer printers and cash registers. The numeric characters used generally represent a fixed-font style and may include certain control symbols in addition to the numerals 0 through 9.
2. *Alphanumeric Sets.* More varied alphanumeric data, such as typewritten text, can be read by many of the current optical character readers. Although most of the characters (except for the special symbols) appear on present printers and typewriters, the variations

in special symbol notation have caused compatibility problems in equipment. Present users of optical character readers have in some cases modified their typewriters and computer printers at a nominal expense in order to gain compatibility.

Most readers are restricted to upper-case letters. Units that can read both upper and lower case are especially useful in applications requiring the processing of field-prepared documents.

FONTS

Even more troublesome than the diversity in character sets is the wide variety of type fonts. Literally hundreds of different type fonts are in fairly widespread use, and more than 100 font styles can be recognized by some of the current multifont readers. At first this diversity might seem relatively unimportant, since an "A" is basically an "A" regardless of the font, and shape differences between fonts do not seem to be extreme. A glance at Figures 3-1 and 3-2 shows how different a single letter might appear in several styles, and how it would be difficult to distinguish a capital I, a lowercase l and the numeral 1 (see Fig. 3-2) when the slight differences in the shapes themselves are compounded by the slight differences in font styles. Human beings have little difficulty in recognizing these letters and numbers largely because the legibility of individual letters or even of individual words is usually not critical when a person reads, since human beings read letters within the context of the entire word and words within the context of the entire sentence. Consequently the word "ouic" in the phrase "ouic and dirty" could be identified in context by most human readers as the word "quick," even though the first letter of the word "ouic" is an "o" and the last letter "k" is missing.

Optical character readers at this point in time must rely on shape; any additional information such as identification of font style or individual numeric field is usually preprogrammed. The more complex multifont and omnifont readers may include recognition logic for automatically determining whether or not the input font is one that resides in memory. But if it is not in memory, it is usually better to avoid erroneous identifications by not reading it at all, unless the device has a program for automatically developing reference recognition logic for new fonts while the character set is read in and identified. Scan Data Corporation, CompuScan, Inc., and Information International have readers with this capability.

Development of context recognition logic in an OCR device is a difficult problem, but attempts to simulate the ability of human beings to read by context have been made in some long-range developmental projects.

The first requirement for automating the process of context recognition is a group of fundamental rules that will aid the machine in identifying

ABCDEFGH
IJKLMNOP
QRSTUVWX
YZ*+,-./
01234567
89

ISO Size B Type Font (lowercase omitted)

ABCDEFGHIJKLM
NOPQRSTUVWXYZ
0123456789
•¬:;=+/$*"&|

CONTROL SYMBOLS

'-{}%?⌐Ч⌐

ÜÑÄØÖÆ£¥

ANSI OCR Size A Type Font (lowercase omitted)

Farrington 12L Type Font *(no lowercase)*

Farrington 7B Type Font
(no alphabetics)

Farrington 12F Type Font
(no alphabetics)

IBM 1428E Type Font
(no alphabetics)

IBM 1428 Type Font
(alphabetics omitted)

NCR Optical Font (NOF)
(no alphabetics)

Fig. 3-1. The common OCR fonts

Note: IBM 1428 is alphanumeric, but only numeric characters are illustrated here.

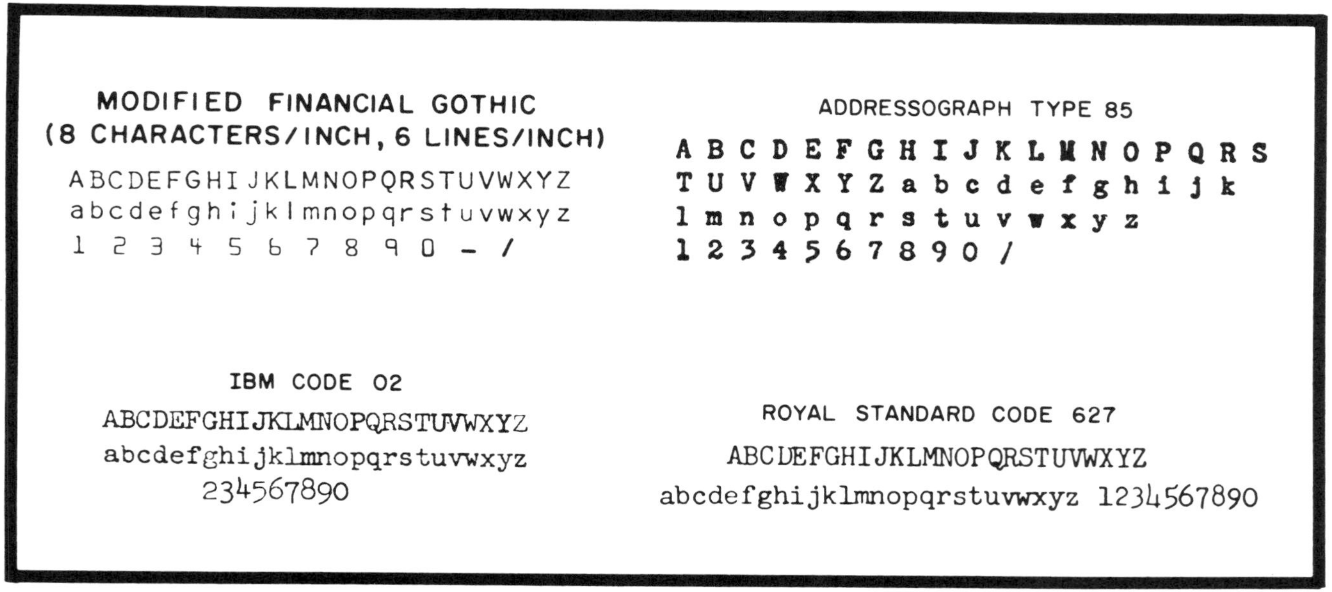

MODIFIED FINANCIAL GOTHIC
(8 CHARACTERS/INCH, 6 LINES/INCH)

ABCDEFGHIJKLMNOPQRSTUVWXYZ
abcdefghijklmnopqrstuvwxyz
1234567890 - /

IBM CODE 02

ABCDEFGHIJKLMNOPQRSTUVWXYZ
abcdefghijklmnopqrstuvwxyz
234567890

ADDRESSOGRAPH TYPE 85

ABCDEFGHIJKLMNOPQRS
TUVWXYZabcdefghijk
lmnopqrstuvwxyz
1234567890/

ROYAL STANDARD CODE 627

ABCDEFGHIJKLMNOPQRSTUVWXYZ
abcdefghijklmnopqrstuvwxyz 1234567890

Fig. 3-2. Typical typewriter font styles.

characters on the basis of the context in which they are used. These context rules must agree with the type of material being read; i.e., if a new application is added, then new rules should be instituted. These rules can be changed by utilizing programming techniques.

Although context recognition is not yet sophisticated enough to become the major element of a recognition scheme, it can be used as a backup method for identifying illegible characters. The most obvious advantage is the ability to identify a complete word even if one or two characters cause recognition difficulties. Context recognition will involve an enormous increase in the storage capacity and logical capabilities of character readers, but this growth may be justified by the gain in efficiency. However, the economics of context-recognition readers will remain highly speculative until considerably more development work has been undertaken.

Context recognition also promises to be useful for reading handwriting; it could be the basis of a technique for reading complete words rather than a character at a time. Again, this approach would radically increase the storage requirements for and the cost of a reader, but the results might justify it. Again, the economics will remain unclear pending additional development work.

At present there are five distinct levels of font-reading capability, governed chiefly by ability to distinguish shapes:

1. Ability to read a single printed or typewritten font.
2. Ability to read a few selected groupings of printed or typewritten fonts, with the reader usually preprogrammed to "expect" a particular font. Some companies call these "multiple-font" readers as distinguished from "multifont" readers. Others make no distinction.
3. Ability to read a variety of fonts and to mix fonts on a given page. Such machines are called multifont readers.
4. Ability to read any font, by a given character-set analysis immediately prior to reading. Such machines are called omnifont readers.
5. Ability to read handwritten or hand-printed characters, usually an auxiliary capability.

Manufacturers of single-font and multifont OCR equipment may claim that most fonts are similar in basic character shapes; therefore their machine can actually recognize more font styles than those specified for it. It should be stressed, however, that since the recognition unit is designed to read only the specified fonts, accuracy cannot be guaranteed for different fonts, no matter how similar; in fact there is likely to be an increase in reject and error rates. A machine like the CompuScan 370, however, is designed to "acquire" fonts immediately before reading the data if necessary, and can therefore be correctly called "omnifont."

One of the problems faced by multifont and omnifont readers is the variation in type sizes as well as type styles. The size of the area that the letter occupies needs to be defined in order to determine all aspects of an individual shape; this is difficult when type sizes vary. This problem is encountered also by readers that must deal with proportional spacing, since again the size of the area containing the character is not predictable. Needless to say, single-font and most multiple-font and multifont readers deal with fixed-pitch (fixed spacing) fonts rather than proportional spacing, and have a standard as to allowable type size in order to provide enough decision points to identify the character accurately.

Figure 3-3 shows a difficult text that can be read only by a highly versatile reader because of the small size, proportional spacing, and range of type fonts employed.

Because of the cost of attempting to recognize characters that closely

---

Exposure: Photography; Standards; *9372.*

Extension: Grain-crack; Mullen test; Burst of leather; *9466.*

Exterior: Grading; Interior; Moisture content; Overlay; Plywood; Shear test; Vacuum and pressure test; Adhesives; Boil test; Bonding; *PSI-66.*

External tensor permeability; Ferrimagnetic resonance; Ferrites; Garnet; Low magnetic field effects; Permeability; Spin waves; Tensor permeability; YIG; *9395.*

External wall units; Innovation; Long-time behavior; Performance standards; Accelerated tests; Bathtubs; *9452.*

Extinction coefficient; Polymer; Polystyrene; Solvent effects; Ultraviolet absorption; Carbon tetrachloride; Cyclohexane; Ethylbenzene; *J. 71A 2-449, 169-175 (1967).*

Extrinsic faults; Silver-tin alloy; Stacking fault energy; Dislocations; *9459.*

F

Fabry-Perot interferometer; Airglow photometry; *J. 70C 3-226, 159-163 (1966).*

Face; Frame; Lineality space; Linear programming; Algorithm; Cone; Convex hull; *J. 71B 1-189, 1-7 (1967).*

Factor set; Group extension; Level; Modular group; Parabolic class number; Quotient surface; Split extension; *9317.*

Fading characteristics; Textiles; Fading standardization; Light-sensitive paper; Booklets of faded strips; *M260-15.*

crystals; Zone refining; *9197.*

Fermi surface; Orientation; Potassium; Spark cutting; Zone refining; Crystal growth; *J. 71A 2-443, 127-132 (1967).*

Ferrimagnetic resonance; Ferrites; Garnet; Low magnetic field effects; Permeability; Spin waves; Tensor permeability; YIG; External tensor permeability; *9395.*

Ferrites; Garnet; Low magnetic field effects; Permeability; Spin waves; Tensor permeability; YIG; External tensor permeability; Ferrimagnetic resonance; *9395.*

Ferro- and ferrimagnetism; Magnetization; Magnetometer calibration; Measurement of magnetization; Saturation magnetization; Vibrating-sample magnetometer; *J. 70C 4-236, 255-262 (1966).*

Ferroelectrics; H_2O; Ice; Inelastic scattering; $K_4Fe(CN)_6 \cdot 3H_2O$; Librations; Low-frequency modes; Neutron cross-sections; Neutron spectra; Phase transition; *9504.*

Ferromagnetism; Hyperfine field; NMR; Nuclear interaction; Nuclear quadrupole resonance; Thermal expansion; *9219.*

Ferrous SRMs; Stoichiometric mixtures uranium oxide; Isotope ratio determination; Spectrophotometric titration; Controlled potential coulometric; Molybdenum; Homogeneous precipitation; Aluminum; Beryllium; Thermoanalytical standards; Tricalcium silicate; *TN402.*

Ferrous standard reference materials; CO by infrared absorbancy; Homogeneity; Certified value; Eighteen cooperating laboratories; Vacuum fusion; Inert gas fusion; Oxygen; *M260-14.*

150

Kryder, S. J., Birnbaum, G., Lyons, H., Microwave measurements of the dielectric properties of gases. J. Appl. Phys. **22**, No. 1, 95 (1951). 259.

Kryder, S. J., Maryott, A. A., Dipole moment of perchloryl fluoride. J. Chem. Phys. **27**, No. 5. 1221 (1957). 2089.

Krynitsky, A. I., Stern, H., Experimental production of nodular graphite in cast iron. Foundry **80**, No. 3, 106; Pt. 2, 80, No.4, 98 (1952). 444.

Kuder, M. L., A broadband, low-level, error-voltage detector. Rev. Sci. Instr. **25**, No. 5, 464 (1954). 914.

Electron-tube curve generator. Electronics **25**, No. 3, 118 (1952). 440.

Kuder, M. L., Hammond, H. K., III, Holford, W. L., Ratio-recording spectroradiometer. J **64C2**, 151 (1960).

Kuehner, E. C., Leslie, R. T., Review of analytical distillation. Anal. Chem. **30**, 629 (1958). 2744.

Kuentzel, L. E., Schoen, L. J., Broida, H. P., Glass dewars for optical studies at low temperatures. Rev. Sci. Instr. **29**, No. 7, 633 (1958). 2575.

Kuhn, H. W., Contractibility and convexity. Proc. Am. Math. Soc. **5**, No. 5, 777 (1954). 978.

Solvability and consistency for linear equations and inequalities. Am. Math. Mo. **63**, 217 (1956). 1869.

Kuhn, H. W., Hoffman, A. J., On systems of distinct representatives. Ann. Math. Studies No. 38 (1956). 1793.

concentrate. J. Electrochem. Soc. **102**, 77 (1955). 1333.

Lagoda, H., Wiener, M., Background eradication in thick-layered emulsions. Rev. Sci. Instr. **21**, No. 1, 39 (1950). 30.

Laki, K., Mandelkern, L., Posner, A. S., Diorio, A. F., Mechanism of contraction in the muscle fibre—ATP system. Proc. Natl. Acad. Sci. **45**, 814 (1959). 2986.

Lalos, G. T., Broida, H. P., Rotational temperature of OH in several flames. J. Chem. Phys. **20**, No. 9, 1466 (1952). 534.

Lalos, G. T., Corruccini, R. J., Broida, H. P., Design and construction of a blackbody and its use in the calibration of a grating spectroradiometer. Rev. Sci. Instr. **29**, 505 (1958). 2513.

Lam, D. G., Jr., Wachtman, J. B., Jr., Young's modulus of various refractory materials as a function of temperature. J. Am. Ceram. Soc. **42**, 254 (1959). 3114.

Lam, D. G., Jr., Wachtman, J. B., Jr., Teffl, W. E., Stinchfield, R. P., Elastic constants of synthetic single crystal corundum at room temperature. J **64A3**, 213 (1960).

Lamb, J. J., George, D. A., Baker, H. A., Sieffert, L. E., Impact strength of some thermosetting plastics at low temperatures. ASTM Bull. No. 181, 67 (1952). 464.

Lamb, J. J., Reinhart, F. W., Boor, L., Brown, C., Evaluation of the Boor-Quartermaster snag tester for coated fabrics and plastic films. ASTM Bull. No.

Fig. 3-3. Two examples of texts read by CompuScan 370.

resemble one another, OCR manufacturers were quick to see the advantages of a highly stylized font designed to provide a number of characteristics that would serve to distinguish each character. An early pioneer in this work was Farrington Electronics, which developed two "Self-Check" fonts, the 7B and 12L fonts (see Fig. 3-4), which provide at least two distinguishing features per character. The NOF numeric font of NCR (see Fig. 3-5) is highly readable and requires only a simple decision matrix. The decision processes for both fonts are illustrated in Figures 3-4 and 3-5. Also de-

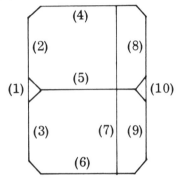

(1) Long vertical left
(2) Short vertical upper left
(3) Short vertical lower left
(4) Horizontal top
(5) Horizontal middle
(6) Horizontal bottom
(7) Long vertical right
(8) Short vertical upper right
(9) Short vertical lower right
(10) Long vertical extreme right

Fig. 3-4.   Farrington character elements for the 7B and 12F fonts.

veloped by IBM is an alphanumeric "1428" font with a "1428E" variation designed to be read by character readers.

In 1966, the American National Standards Institute, Inc. (ANSI), under the sponsorship of the Business Equipment Manufacturers Association, completed a thorough study of OCR requirements and adopted a standard character set that, while not unanimously agreed upon, was generally accepted by the manufacturers of OCR equipment. The Standard specified

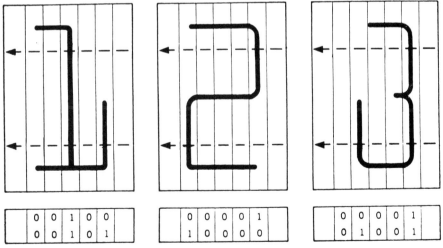

Fig. 3-5.   Recognition points and resulting internal bit codes of NCR optical font (NOF).

three sets (commonly called OCR-A, OCR-B, and OCR-C), which are identical except with respect to overall size. The set OCR-A (listed as X3.17-1966 in the 1971 ANSI catalog) facilitates recognition to such a degree that there is a substantial trend to produce lower-cost, single-font or double-font (OCR-A plus one other) readers, particularly in the category of document readers. Formerly, OCR-A included only upper-case alphabetic characters, but standards were also adopted in 1970 for lower-case characters. Printers capable of both upper- and lower-case OCR-A and relatively inexpensive single-font readers capable of recognizing the entire set should make an appearance by mid-1971.

During the development phase of the OCR project by ANSI, the world-wide parent standards agency, the International Standards Organization (ISO), adopted its own "style B" standard character set. Note that the two standards differ significantly. Although the ISO standard is certainly more esthetically pleasing, American manufacturers feel that it is much less stylized and therefore is less readily recognizable by OCR equipment than the ANSI set. Actually, a number of American manufacturers who have worked with this font say that the ISO set is just as machine-readable as the ANSI font, and has the advantage of looking acceptable for conservative business correspondence. Others like the "modern" look of the American font. The ISO set (listed as ISO standard R1073-1969 in the 1971 ANSI catalog), sometimes also called OCR-B, has included lower-case alphabetic characters for some time, but has not achieved widespread use in this country. The ANSI set was recently expanded to include lower case; a proposal to endorse the American OCR-A font as a second international "Style A" ISO character set is currently being considered and will probably be adopted.

RECOGNITION OF HANDWRITING

Since each individual has his own style of handwriting, it is difficult to set recognition standards that will not lead to a high reject rate. Consequently, this problem is even more perplexing than the multifont recognition problem because the recognition logic of the machine can never be set for a particular mode.

The work being done on the recognition of handwritten characters can be divided into two classes: hand-printed characters and script. Some of the techniques being investigated in connection with script are curve tracing, detection of selected features, and context recognition. Although a number of companies are working on the problem, most of the work has been kept confidential. The primary customer for a reader capable of handling script appears to be the U.S. Post Office Department.

There is definitely a widespread commercial use of hand-print recognition, however, particularly for the entry of source-data records and bill-

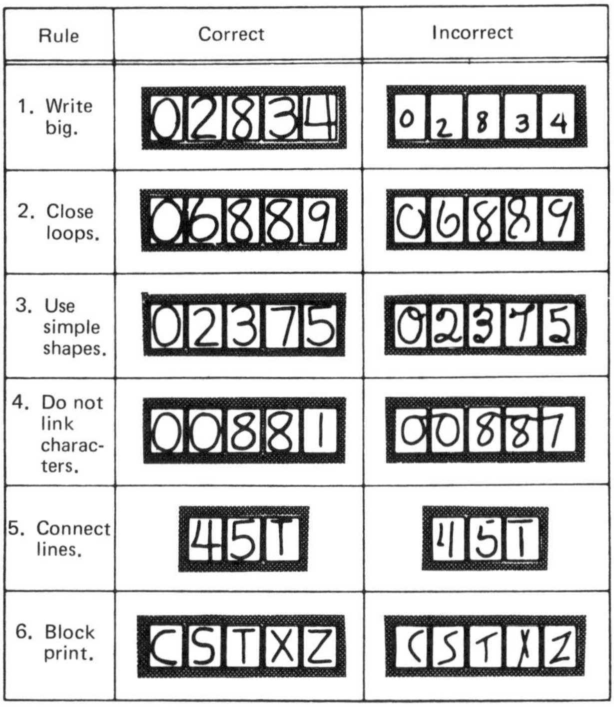

Rule	Correct	Incorrect
1. Write big.	0 2 8 3 4	0 2 8 3 4
2. Close loops.	0 6 8 8 9	0 6 8 8 9
3. Use simple shapes.	0 2 3 7 5	0 2 3 7 5
4. Do not link characters.	0 0 8 8 1	0 0 8 8 1
5. Connect lines.	4 5 T	4 5 T
6. Block print.	C S T X Z	C S T X Z

Fig. 3-6.   IBM 1287 handwriting rules.

ing exceptions. The ıʙм 1287 and 1288 optical readers can read both hand-printed numeric and selected alphabetic symbols, but a glimpse at the rigid set of rules for numerics (illustrated in Fig. 3-6) indicates that the concept is still quite restricted in scope. However, a number of other readers with hand-print capabilities, including the Recognition Equipment Input II system and the Control Data 936-1, have recently appeared, attesting to the progress in this field.

The reject rate in hand-print systems can be considerably affected by a training program for those personnel who will be generating input documents. Training programs in both Germany and Japan have had excellent results in lowering the reject rate; similar programs in this country have also had very good results, although not so good as those abroad.

International Business Machines, a pioneer in the research on hand-printing, now offers an upper-case alphanumeric hand-print option on the 1287 and 1288 readers. Other manufacturers may or may not follow suit, since the problems of recognition are considerably more difficult with the larger character set. Some manufacturers such as Optical Scanning Corporation (see Fig. 3-7) have experimented with a preprinted pattern that guides the person in printing the correctly shaped characters. Undoubtedly, with the widespread interest in this capability, the error and rejection rates for numeric hand-printing will go down as manufacturers improve their recognition techniques.

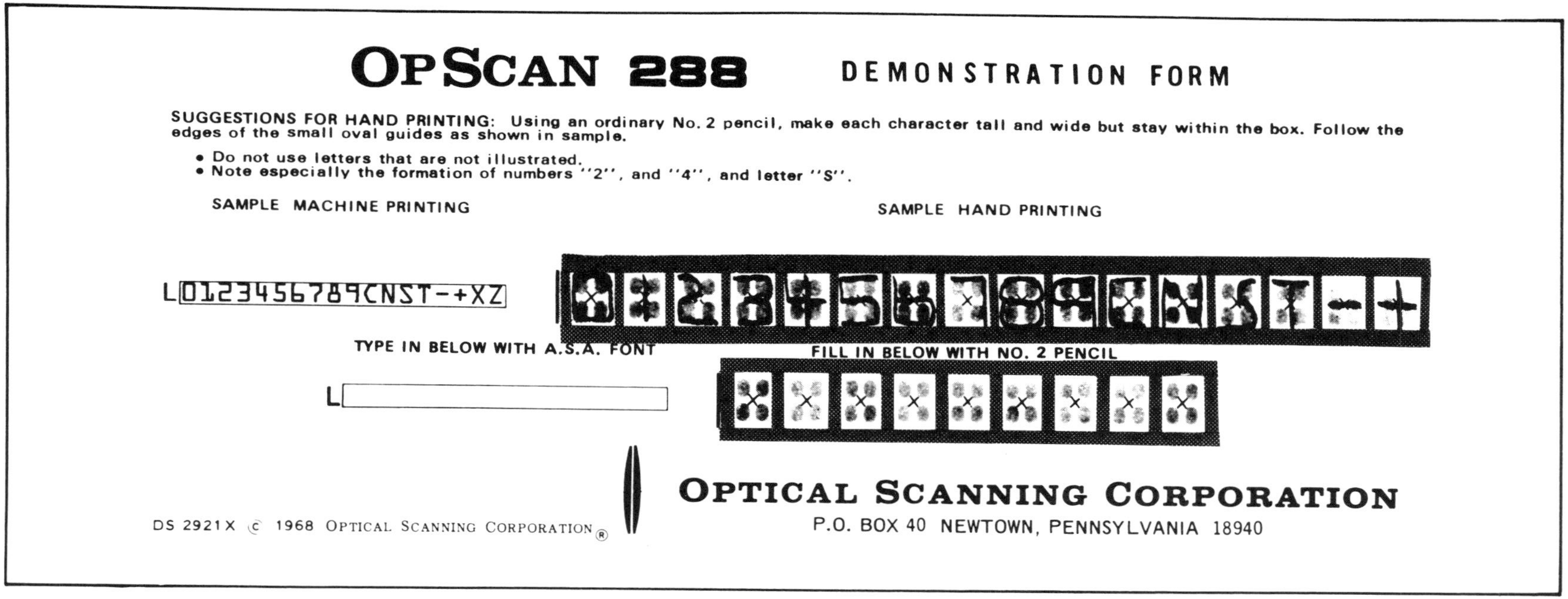

Fig. 3-7. Standard character set for the OpScan 288 and suggestions for hand printing (the letters X and Z can also be hand-printed).

36

4. INPUT PROBLEMS II:
PRINTERS, INK, PAPERS, AND FORMS DESIGN

The fact that many OCR devices deal to one degree or another with source-data recording further complicates the problem of recognition presented by the variety of fonts in current use. Since OCR recognition identifies the character by analyzing it in its relation to its contrasting background, the addition of creases, holes, or dirt marks that alter the reflectance of the white background will in effect change the character shapes of the letters. Most people are aware of the punched-card injunction, "Do not fold, spindle, or multilate," and most people exercise care in handling MICR-coded checks. But OCR forms are not standardized and probably will never be, so that it will always be difficult to try to train the world at large not to fold, spindle, or multilate OCR turnaround documents. Attempts to avoid problems by using heavy-weight paper resistant to wear are only partly successful, since the paper, regardless of weight, may be folded or stapled, or carried around in pockets and purses. Users planning to use OCR forms for in-house data collection may save time and money by training personnel to avoid the type of handling that inevitably occurs when OCR forms go out to the public at large.

Next to the human paper handler, the preparation element most likely to degrade OCR input is the original printing device itself. The OCR devices are increasingly expensive in proportion to their ability to handle mis-registered characters, skewed lines, and blurred characters. The generation of a precisely printed original has a great deal to do with the cost of a reader of sufficient quality to achieve the desired throughput.

Printing for OCR comes from three sources:

1. Typesetting and offset printing machines used by preprinted forms manufacturers or other printing houses
2. Source data typewriters, and
3. Computer printers

Typesetting machines used by printing houses can usually produce clear, well-registered characters, but offset methods of printing often compound faulty printing, as shown in Figure 4-1. Source-data typewriters, whether the hammer type or the revolving printing element, are also capable of producing good OCR input, provided the machines are kept in good

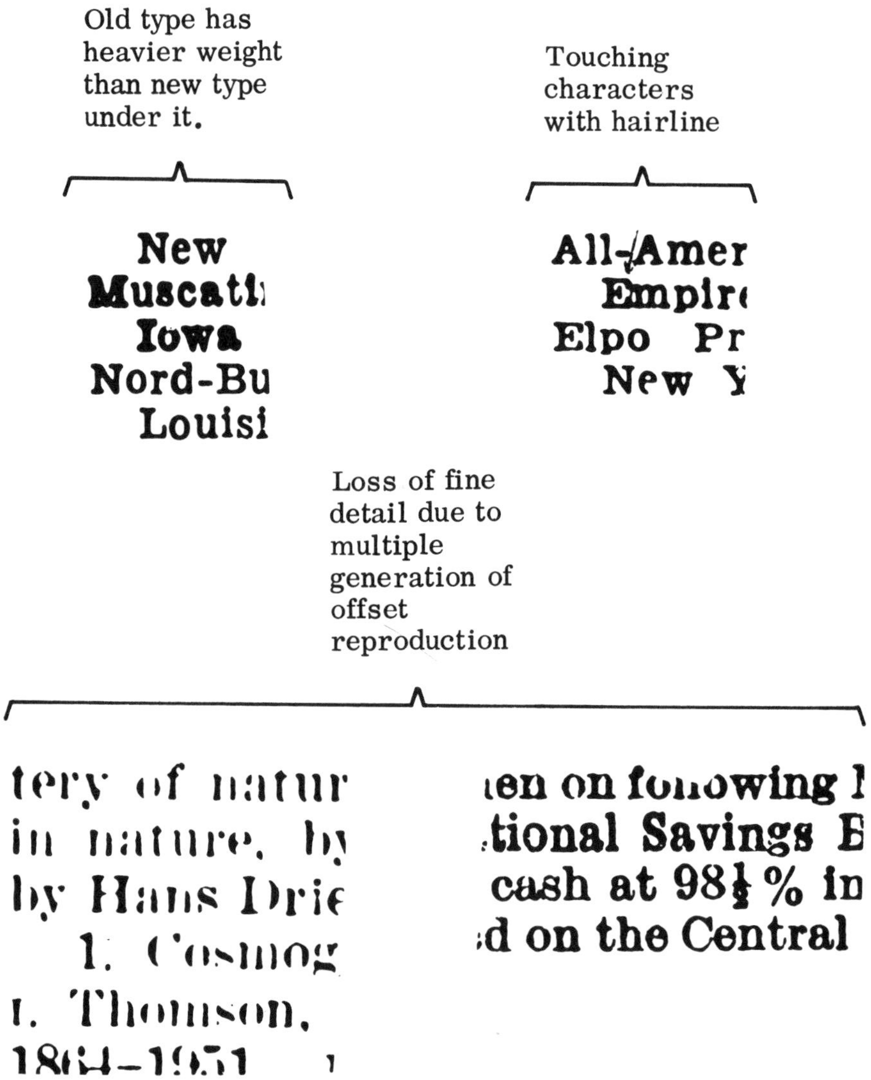

Old type has
heavy weight
than new type
under it.

Touching
characters
with hairline

New
Muscatl
Iowa
Nord-Bu
Louisi

All-Amer
Empir
Elpo Pr
New Y

Loss of fine
detail due to
multiple
generation of
offset
reproduction

tery of natur
in nature, by
by Hans Drie
1. Cosmog
1. Thomson,
1864–1951

en on following
tional Savings E
cash at 98½% in
d on the Central

Fig. 4-1. Common printing defects currently being analyzed with Compu-Scan 370.

working order, the type bars or printing elements are replaced as soon as they begin to produce degraded images, and the ribbons are high-quality "ocr" ribbons. The high speed of computer outprint printers and the resulting image degradation, however, present special problems to the ocr user. These are problems of considerable importance, since he will frequently generate turnaround documents or forms on his computer and, after suitable entries in the field, read them into the system again.

There are basically two main classes of high-speed computer printers available on the market today: impact printers and nonimpact printers. The impact printers are the older class. These print by striking a character image against the paper, allowing a master and a number of copies to be created simultaneously. These printers are, of necessity, mechanical devices and by nature have an upper limit on speed because they are restricted by the mechanical nature of the movement required. Most modern impact printers use an "on-the-fly" approach in which high print speeds are achieved by extremely rapid hammer action against continuously moving type elements. The principal variations involve the use of a rotating drum, a horizontally moving chain, or an oscillating bar, as detailed below; the actual methods of printing are quite similar in all three techniques.

IMPACT PRINTING METHODS

During each print cycle (normally the time allocated to load the print buffer, decode its contents, print one line including hammer action and recoil, and space the paper), all characters move past the print hammers at each printing position. The character to be printed is selected by decoding, and a fast-action hammer controlled by an actuator presses the paper against the type slug at the exact moment the required character is in position. If the machine is printing at 600 lines per minute, each total printing cycle takes one six-hundredth of a minute. This interval is in turn divided into discrete timing units for each character that is available, plus several units for paper advance.

In the asynchronous mode of printing the firing of the hammers does not commence at any fixed point during the rotation of the character set. Rather, firing commences whenever a signal is received to indicate that line spacing has been completed and the print buffer loading is finished. Firing terminates when a counter indicates that all characters have moved past the hammers or when the buffer holding the line of characters to be printed has been sensed and found empty.

Hammer action in "on-the-fly" printers is either by (1) free flight, or "ballistic," hammers (movement stopped by contact with the paper and the type element); or (2) "controlled flight" hammers (fixed spatial movement). The most important advantage claimed for the latter design principle is positive control over the depth of penetration of hammer action. When such a printer is operated without paper in the tractor feed, the hammers are prevented from striking the type element by "end of paper" safety switches.

Vertical format control is generally effected by an 8- or 12-channel paper-tape loop. The vertical spacing of the punches controls the actual spacing on the printed sheet. In some printers it is necessary to use a loop that has the exact vertical size of the printed page; in others it is possible to use loops representing only the vertical area to be imprinted. It is usually possible to space the printer under program control.

The general characteristics of current "on-the-fly" printers are:

high speeds (300 to 1200 or more lines per minute)
the absence of a platen
ribbon movement parallel with paper motion; ribbon width at least
    equal to maximum line length
hammers that strike the paper from behind.

There are three basic kinds of "on-the-fly" printers: drum printers, chain printers, and oscillating-bar printers. These vary according to printing techniques, as explained below.

Drum Printers

Examples of drum printers are the Honeywell 222 and the GE PRT 201. A widely used on-the-fly printing technique is to provide a complete character set (sometimes two or more complete sets) at each print position, and to distribute these character sets around the circumference of a solid, continuously rotating drum. The timing mechanism senses the passage of a particular character in front of the hammers and then fires the hammers that correspond to the positions in which the given character is to be printed. Thus, if all the hammers were fired at the same instant, the printed line would consist of the same character printed at all positions.

Several drum printers have utilized the "shuttle" technique, which cuts in half the number of hammers needed and hence reduces the cost. In a "shuttle printer" (e.g., Anelex 4000), the odd-numbered columns are printed in one cycle; then the paper is "shuttled" one column to the left and the even-numbered columns are printed. Note that since two cycles are needed to print each line, the effective speed is halved.

Chain Printers

Examples of chain printers are the IBM 1403, Potter HSP-3502, and CDC 512. In a chain printer the hammers must be individually timed because each character travels horizontally across many printing positions during the print cycle. Several identical sets of characters are assembled serially on a horizontally moving chain, which resembles a bicycle chain. At each print position the paper is forced into contact with the ribbon against the chain by a solenoid-activated hammer fired as the appropriate character on the chain passes the printing position. In the IBM 1403 Model 3, the chain has been replaced by a "train" mechanism in which type slugs move in the same horizontal plane as in the chain, but at more than twice the speed of the original 1403 printer. If all hammers were fired simultaneously in a chain printer, several sets of sequential characters rather than a line of identical characters would be printed. (Of course, the print line would be distorted if this were done, since normally only one-third of the print hammers are aligned with the slugs at any one time.)

Oscillating-Bar Printers (OBP)

Examples of the bar printers are IBM 1443, UNIVAC 3030, and Datamark OBP. An oscillating-bar printer operates much like a chain printer except that the print slugs are inserted in a horizontal bar that moves rapidly back and forth instead of being attached to a continuously traveling chain.

The highest printing speeds that can be achieved using this start-stop-reverse type of motion are considerably lower than those that are possible with a continuously rotating chain or drum; the fastest available oscillating-bar printer operates at about 600 alphanumeric lines per minute. However, a bar printer is likely to cost less than a drum or chain unit of comparable speed, and it offers the added advantage of permitting rapid removal and replacement of type bars, a valuable asset where an installation's application mix requires the use of several different character sets.

NONIMPACT PRINTING METHODS

A breakthrough in printer design has come in the form of nonimpact printing techniques such as photographic, xerographic, or cathode-ray-tube methods. Several nonimpact printers are now on the market. One uses an interesting technique in which a character is written on a CRT and then piped through a fiber-optics cord to print on light-sensitive paper; this unit is reported to run at 6000 alphanumeric lines per minute, and is an example of the sort of inspired design of which the industry is capable. A second idea is to shoot electrons at specially treated paper; a third is to charge the ink particles themselves and shoot them at the paper in a stream that is electronically deflected into character patterns.

The major hurdles facing nonimpact printer manufacturers are (1) that most nonimpact printers require specially treated paper, which is expensive and often has a consistency that is unpleasant to touch; and (2) that these printers are, at present, incapable of producing more than one copy of the printout, which is a crippling disadvantage in some commercial applications. It is interesting to speculate that perhaps nonimpact printing devices will eventually be so inexpensive as to allow the purchase of multiple units, all driven in parallel by a single set of electronic logic; yet, even there, the question of whether such parallel-produced documents are legal copies of one another will have to be resolved.

Thus, nonimpact printing techniques offer the potential for high printing speeds at comparatively low costs, but some serious problems will have to be overcome before they can become effective across-the-board competitors with mechanical printers. The impact printer is here to stay, and while it will be supplemented in certain applications by the newcomers, it is not likely that it will soon be supplanted by them.

COMPUTER PRINTERS AND OCR

Each of the various printing techniques now available has its own problems when utilized as a means of printing for OCR.

While "on-the-fly" impact printers have been characterized from their inception by frequent misalignment or misregistration of the printed characters, it may be safely stated that printing quality has been greatly improved, especially on the more expensive printers. Of the two types of printers, the chain printer probably produces the best images for OCR recognition. Any misregistration in chain-printed copy is horizontal; that is, the spacing between adjacent characters is uneven. This type of misregistration is less noticeable than waviness of the printed line, which is characteristic of drum printers. Even the latter are, in general, now capable of producing high-quality copy with little detectable waviness when properly adjusted and maintained. Both types need to be used with high-quality "OCR" ribbons to provide an even reflectance and sufficient weight for trouble-free feeding and transport.

Some nonimpact printers can also generate printed pages suitable for OCR input. Two problems tend to be common to nonimpact printers, the type of paper used and the sharpness of the character image. Many of the treated papers used with nonimpact printers are not heavy enough to be fed to an OCR device without jamming it. Although most nonimpact printing methods are capable of printing lines equal to or better than impact printers as far as registration is concerned, many techniques produce characters that have rather blurred outlines and which lack the precision in shape ideal for OCR.

INKS AND PAPERS

The use of special ribbons and papers for OCR are important in order to produce sharp images resistant to deterioration. Scanners vary with respect to the kind of light used to scan images, so that all "black" inks are not equally black to different scanners. The ink and ribbons used with OCR have to be carefully chosen in order to provide a minimum of reflectance, and the papers must be chosen so that the reflectance from white paper is not uneven. In addition, if preprinted forms are to be used, the use of special "nonread" ink geared to the scanner's light will often permit more efficient processing. In most cases the manufacturer knows which suppliers produce the type of paper, printing ink, and nonread ink most suitable for his particular scanner.

FORMS DESIGN

In many cases when data are gathered from the field it is possible to exercise some control over the format of the source data input through design of a form that can be read more efficiently. The OCR forms design is concerned not only with maintaining margins and read areas compatible with the scanner (see Fig. 4-2), but also with placement of the data in pre-defined locations that require a minimum of searching. Two examples of U.S. government forms used to produce typed documents are shown in Figures 4-3 and 4-4. The first, a previously existing design, exerts little control over data placement. A minimum set of guidelines is provided; although the form can be read optically, it provides a real challenge to the equipment and a gamble to the user with respect to data accuracy. The scanning field is so large that the OCR reader cannot operate at its maximum speed. Only the more expensive multifont/multiline OCR machines can process such documents, and it is usually difficult to justify the economic feasibility of reading documents such as these by optical means.

A much better approach appears in Figure 4-4. This form has been designed specifically for use with optical character readers. Data fields are of restricted block-entry design, and instructions or other markings in juxtaposition with the entry fields are printed in nonread ink (shown as gray-shaded areas in Fig. 4-4) invisible to optical scanning devices. This practice eliminates the possibility of the reader's picking up "noise" clutter.

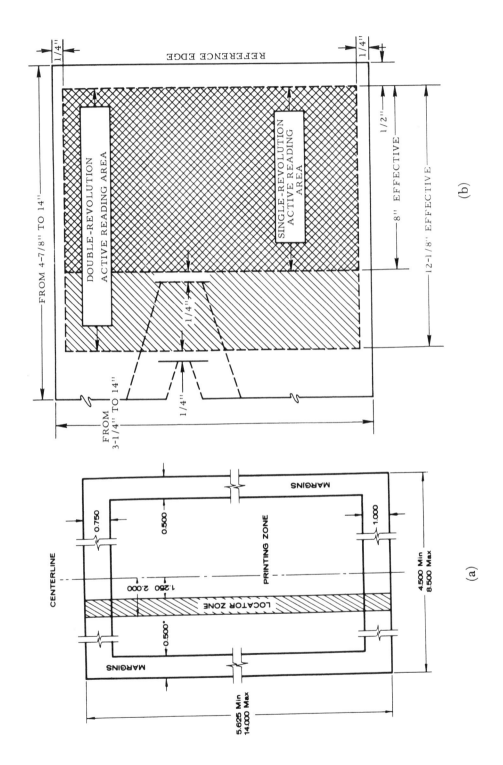

REFERENCE EDGE

FROM 4-7/8" TO 14"

DOUBLE-REVOLUTION
ACTIVE READING AREA

SINGLE-REVOLUTION
ACTIVE READING
AREA

1/4"

1/4"

1/2"

8" EFFECTIVE

12-1/8" EFFECTIVE

1/4"

FROM
3-1/4" TO 14"

1/4"

(b)

MARGINS

MARGINS

CENTERLINE

0.750

0.500

PRINTING ZONE

1.000

LOCATOR ZONE

1.250 2.000

0.500°

2.000

0.500°

4.500 Min
8.500 Max

5.625 Min
14.000 Max

(a)

44

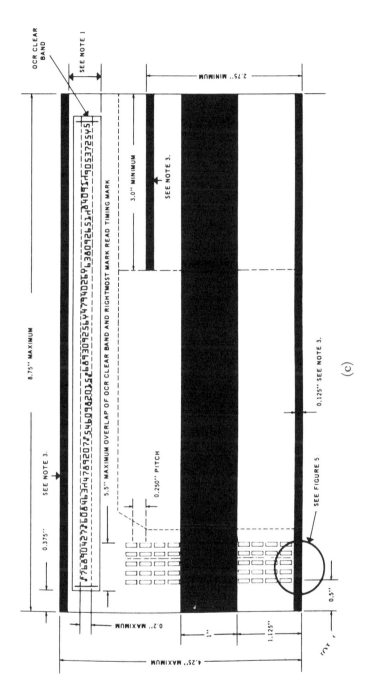

Fig. 4-2. Examples of form design requirements for OCR devices: (a) Farrington 3050; (b) Recognition Equipment ERCR; (c) UNIVAC 2703.

45

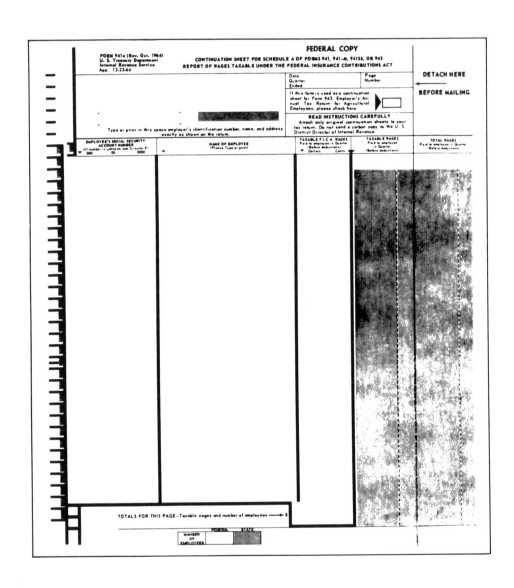

Fig. 4-3. A previously existing form adapted to OCR use.

46

OE 2300-5.2
HIGHER EDUCATION GENERAL
INFORMATION SURVEY
PAGE 4

I

U. S. DEPARTMENT OF HEALTH,
EDUCATION AND WELFARE
OFFICE OF EDUCATION
WASHINGTON, D. C. 20202

09   03

Analysis of resources

For your present total collection (sum of items 4-7 part A) enter in column 4 below the approximate percents which are devoted to the areas in column 1. Likewise, enter in column 5 the approximate percents for your current acquisitions (item 2 part A). In each column the percents should total 100. (Provide approximations by measuring your shelf list cards at 100 cards per inch; estimate for unclassified serials or periodicals. The numbers and letters below refer to the approximate D.C. and L.C. schedules.)

8	AREA	CLASSIFICATION		% OF TOTAL COLLECTION	% OF CURRENT ACQUISITIONS
		D.C.	L.C.		
9		000, 100, 200, 400, 700, 800	A, B, M, N, P, Z		
10		300, 900	C, D, E, F, G, H, J, K*, L		
11		500 - 559	Q, QE		
12		560 - 599 610 - 619	QH - QR R, S		
13		600 - 609 620 - 699	T, U, V		
14					

*r.e. LAW (pending)

PART II - LIBRARY FACILITIES

Assistance in completing this section should be obtained from the Director of Physical Plant or the Chief Business Officer. Provide estimated or approximated data, if necessary. Floor space in square feet is defined for this item as the area of a room based upon measurements taken from inside walls at floor level. Where there are minor architectural projections (less than 12 inches), the distance between typical walls is used. Include areas covered by built-in equipment, such as counters and shelving.

9 15   Total floor space in square feet allocated to library functions in all buildings (including branch and extension center libraries)

16   Of the total area reported in item 15 above, estimate the square feet devoted to each of the following functions:

   a. Stack areas for shelving volumes (include book storage areas)

   b. Seating areas

   c. Staff offices and work areas

   d. Other areas (lounges, exhibits, non book storage, corridors, stairs, etc.)

PART III - LIBRARY MECHANIZATION

Type an "X" where each of the following operations is presently mechanized (i.e., whether alphabetic or alpha numeric information is handled by machine) in your MAIN library.

10	ITEM	TYPE X FOR ONE		IF NO, ENTER FISCAL YEAR IF ANY FOR WHICH MECHANIZATION IS PLANNED
		YES	NO	
17	ACQUISITION			
18	SERIAL RECORD			
19	CIRCULATION			
20	OTHER Specify			

11   PART IV - STUDENT AND OTHER HOURLY ASSISTANCE

21   NUMBER OF HOURS OF STUDENT ASSISTANCE DURING YEAR

22   NUMBER OF HOURS OF OTHER HOURLY ASSISTANCE DURING YEAR

COLLEGE AND UNIVERSITY LIBRARY RESOURCES AND FACILITIES, 1965-66

Fig. 4-4. A controlled-design document.

A number of ocr paper manufacturers and suppliers have been devoting much time and money to developing good forms of this type.

The most important considerations in designing forms for ocr include

1. Location and procedures for document preparation. The printing ribbon or ink, the type of paper (weight, reflectance, color), the carbons, condition of the printer type faces, registration, and skew are all factors that must be interrelated.
2. Specific machine limitations of the reader to be used, particularly with regard to font capabilities, registration tolerances, and skew are factors to be scanned.
3. Separation of ocr fields from non-ocr fields to cut down unnecessary scanning.
4. Minimizing white space between ocr fields to reduce unnecessary scanning.

It can be seen in Figure 4-4 that the data entry blocks are quite different from those of a standard form. All data guidelines or boxes are printed with a blue ink that is not readable by ocr equipment, and the more critical quantity data fields are separated by healthy margins to minimize errors that could result from line skew or misaligned vertical placement.

Learning about in-house applications generally requires approximately a week of training for the typist who will prepare the ocr-readable documents. The purpose of this training is to increase numerical typing speed, to decrease typing errors, and to teach the ocr control symbols used to format the data.

Document preprocessing at the ocr installation consists of removing staples, clips, etc., from the documents; sorting the machine-readable from the non-readable documents prior to ocr processing; separating documents according to size or data contents; and batching the documents. Batching usually involves use of a header sheet that describes document source, contents, and number of documents in a batch. Most ocr devices can also check to see that the proper totals have been read.

Examples of documents field-typed on preprinted forms presently being optically read include government information survey and quarterly tax forms, trucking waybills, and loan applications. Machine reject rates as reported by users are approximately ten times higher for field-typed documents than for computer-printed turnaround documents. Of course the turnaround bill—where the bill is computer-printed, mailed to the customer, and returned with his payment—has lower reject rates because only exceptions (such as when the customer does not remit the exact amount of his bill) are encoded, often in mark-sense or hand-print fields as well as typed fields.

At present, only the very largest of applications can justify the expense of a reader for completely uncontrolled-field documents. These documents exert little or no control over the printing location and/or the font style used. An excellent example of such a group of documents is the plethora of envelopes that must be sorted by the U.S. Post Office. Among these envelopes is every conceivable OCR pitfall, including unreadable type fonts, interference and address degradation, color and texture problems, folds and creases, skew, and improper locations of the data.

Document transmission to the OCR installation through the mail, though inexpensive, is time consuming and can result in loss or damage. Every effort should be made to protect the physical condition of the document prior to OCR input, to increase reading reliability. If the original documents are not physically required by record-keeping laws, alternative means of providing these data for computer input might be to use either digital transmission over telephone lines or facsimile transmission if diagrams are an integral part of the required data. A number of remote OCR systems have been announced in the past year as a response to this problem. It is conceivable, in view of the recent advances in microfilm technology, that microfilm will play an important role in document transmission in the future.

5. DOCUMENT READERS

The document reader, with the clearly defined "turnaround billing" application associated with its development, has been one of the most successful types of OCR devices. As a result, a number of the smaller companies are producing new document readers with simpler transport and recognition problems. With the exception of Scan-Optics, Inc., new page readers have been confined exclusively to the larger, well-established firms. While some document readers delivered in the 1961–1966 period are currently still being marketed, the lower prices and greater capabilities of many of the newer document readers will probably drive the older models off the market in the near future, creating in effect a new generation of document readers.

The definite trend toward simplification affects most, but not all, document readers. The number of companies offering lower-speed devices that read one or two lines of data printed in a single font now outnumber those companies offering readers that handle multifont systems capable of reading five or six lines of data. The more versatile devices are the older of the two types, and companies like Farrington Business Machines, Recognition Equipment, Inc., and IBM continue to offer versatile expensive readers in line with past developments. In the light of past developments in this field, the current offerings of each of these three companies are described below, followed by descriptions of readers marketed by Honeywell Information Systems, Inc., UNIVAC Division of Sperry Rand Corporation, and others new to the document-reader market.

FARRINGTON: THE 3010

In 1955 Farrington first marketed an optical character reader that read 180 cards per minute, printed in one of several fonts. This was one of the first OCR devices on the market, developed by David Shepherd and the Intelligent Machines Corporation. Subsequent models have incorporated increased speed and format flexibility, alphabetic character recognition, and mark-scanning capability. Farrington has relied heavily on stroke analysis in the recognition logic of its optical character and has supported this emphasis with the development of the nine-stroke Farrington Self-Check 12F, 12L, and 7B fonts. The 3000 Series readers all employ mechanical disk scanners that achieve a rated reading speed of 300 to 400 characters per second.

The current Farrington 3010 document reader, shown in Figure 5-1, was first delivered in 1965. It reads card or paper documents printed in any one of the five following standard fonts: the Farrington Self-Check 12F, 12L, or 7B; the IBM 1428 or the American National Standard optical fonts (ANSI OCR), size A. If the user desires, additional font-reading capabilities can be provided. These fonts, which can include both numeric and alpha-

Fig. 5-1. Farrington 3010 document reader.

betic characters, special symbols, and punctuation marks, are illustrated in Figure 3-1. The input consists of data from an accounting machine, typewriter, or other printing device, as well as a turnaround document produced by a high-speed printer. Although the 3010 reader usually deals with tab cards or paper stock of one to five lines in content, it can also accommodate the larger printed areas of documents up to 8.5 inches long and 6.0 in. high. The font capabilities and the basic features of the 3010 document reader qualify it especially for predominantly numeric applications that require alphanumeric encoding as well. Features include right-margin justification, check-digit calculation, three-output stacker sorting, a batch header feature, flagging functions, mark-guide recognition, matrix-mark scanning, a lister, an optional accumulator, and a program plugboard.

A mechanical disk scanner with oscillating mirrors like those of the document reader uses both a rotating disk with a number of apertures and a fixed aperture plate to slice the character image into vertical strips that are projected onto a photosensitive device. Variations in the light intensity are measured to determine the presence or absence of character strokes. "Yes/no" logic circuitry completes the identification of the character on the basis of the presence and absence of strokes. These circuits then convert the decision to a six-bit-plus-parity code and transfers to the buffer the binary code designating the character. Documents containing unidentifiable characters are sorted into an alternate stacker.

The mechanical disk scanner used by several Farrington readers is slower and less flexible than the flying-spot scanners and photocell scanners used by most other companies. Farrington experienced financial difficulty in 1970, perhaps as a result of remaining too long with the slower technology. However, the 4040 journal tape reader delivered in 1969 utilizes a flying-spot scanner, and it is possible that new devices are under development.

RECOGNITION EQUIPMENT, INC.—THE ELECTRONIC RETINA® COMPUTING READER (ERCR)

In 1964 Recognition Equipment, Inc. (REI), delivered the first of its highly versatile and costly ERCR readers. The Recognition Equipment Electronic Retina® computing reader (see Fig. 5-2) with document carrier and/or rapid index page reader is a general-purpose, program-controlled optical reader that reads data from all or part of printed paper pages, documents, or cards at a rated speed of up to 2400 characters per second. The word "retina" in the trade name of the reader is derived from the use of a photosensor array that simultaneously reads an entire character at once. The

Fig. 5-2. REI Electronic Retina® computing reader modules: (a) rapid index page carrier; (b) programmed controller; (c) magnetic-tape transport and controller; (d) line printer; (e) document carrier input unit.

reader can recognize a wide variety of fixed-pitch type styles, which may vary from form to form within the same batch and from field to field within the same form; rescan capability is provided for unreadable characters. The reader can also record data on all or any of the following output media: magnetic tape, paper tape, punched cards, and line printer forms. A data communications interface is also available. The input may be original data from a typewriter or other printing device, and/or a turnaround (reentry) document produced by a computer system. The magnetic-core memory allows extensive editing and format control.

The document carrier (DC) input unit is available for use with the Electronic Retina® computing reader (ERCR), either by itself or with the rapid index page carrier (RIPC). The document carrier, which is illustrated in Figure 5-2, reads one or two lines in a single pass at up to 1200 characters per second. The ERCR with rapid index page carrier can optically read all or part of paper documents typed or printed in virtually any fixed-pitch type style, including its standard 1403 type style. Type styles may vary from form to form within a batch and from field to field within a form. Documents and pages are held on a rotating drum for reading. Normally, one line is read per revolution, and the drum is moved so that the next line can be read.

All or part of the data read from each document can be encoded on it for further sorting by Recognition Equipment's high-speed reader/sorter. The data are encoded in bar-code form on the front or back of the documents by the ink-jet printer at densities of up to 7.5 characters per inch, without slowing the processing rate of the document carrier. The printer discharges up to 48,000 tiny drops of ink per second from a nozzle. Varying electric charges are applied to the drops, which are deflected for printing as they pass through an electric field.

In 1970 REI tried to upgrade the ERCR by announcing a new high-performance document transport called Input 2. In addition it announced a completely new page reading system, Input 80, and an OCR terminal, Input 3. REI, which currently brings in about one-third of OCR revenues, has obviously been actively continuing research into new OCR technologies.

REI: INPUT 2

Input 2 is a lower-cost document reading system that includes Recognition Equipment's Electronic Retina® computing reader (ERCR) for recognition and processing. The Input 2 system reads character sets of 40 or more different machine-printed or 17 hand-printed characters in a variety of fonts and in pitches ranging from 6 to 12 characters per inch for printed material and 4 characters per inch for hand print. One or two lines of data are read in one pass at 600 documents per minute, with a character-reading rate equivalent of 2400 characters per second. Several type faces may be read from a single document, and document sizes may be intermixed within a batch. All type styles, reading tolerances, formatting, flexibility, and computing powers of the ERCR apply to the Input 2 system.

The standard system configuration consists of the Input 2 transport with three output pockets, the ERCR recognition unit with its associated controlling computer, a Selectric console typewriter, a magnetic-tape unit, and a 600-line-per-minute line printer.

Up to nine extra pockets, a substitute 1000-line-per-minute printer, additional fonts or characters in groups of ten, additional core in 4K increments, and three more magnetic-tape handlers may be added to the system. An endorser option permits printing of six numbers on each document read. The Input 2 system may also utilize the new optional "input image" microfilming attachment available with the standard ERCR. The input image system microfilms one or both sides of input documents on 1200 or 1600 daylight-loading 16-mm film cartridges without reduction of processing speed, and employs both a warning system and automatic shut-off when film is exhausted. Auxiliary processor, retrievers, reader-printers, and duplicators for the film are also available.

The Input 2 system is designed as a more economical system than the standard ERCR, and as such has a number of limitations on system configuration that are not imposed on the more expensive reader. The page carrier, ink-jet printer, and multispeed feeder cannot be included in the system, and character or font recognition masks are restricted to a 120 total. The simplified and up-to-date design of the document handler, however, is expected to provide even higher reliability than that of the ERCR.

INTERNATIONAL BUSINESS MACHINES: 1418, 1428, AND 1287 DOCUMENT READERS

International Business Machines is one of the big manufacturers most actively engaged in development of OCR. Both the 1418 numeric reader first delivered in 1961 and the 1428 alphanumeric reader delivered in 1962 utilized mechanical disk scanners like those used by Farrington. Both readers were on-line to the 1400 series and could handle documents printed in a single IBM font at the rate of about 400 documents per minute (500 characters per second).

In 1968 the delivery of the first 1287 with its high-speed, multifont capabilities and hand-print recognition was a significant event in the history of OCR. The 1287 is an on-line input device that reads multifont data from source documents into an IBM System/360 Model 25, 30, 40, or 50. Optional features allow hand-printed, machine-printed, imprinted, and optical-mark fields to be intermixed as the user desires, subject to the restrictions described in Table 5-1. All models read multiple lines of numeric data in the optional Farrington 7B and NCR NOF fonts as well as the standard 1428 and ANSI OCR-A fonts; Models 3 and 4 can also handle the ANSI OCR-A alphanumeric character set. Models 2 and 4 have the added capability of reading journal (cash register, adding machine) tapes. Input document size may vary from a maximum of 4.91 x 9 inches to a minimum of 2.25 x 3 inches. Tapes with widths of 1.31 to 4.5 inches and lengths of 3 to 200 feet can be read by Models 2 and 4 (4.5-inch wide tapes may have a maximum length of 175 feet.)

The 1287 is a synchronous, unbuffered device; once it begins to send

Table 5-1. Acceptable 1285/1287 Characters.

1287 Standard Font		With #5300 on 1287		With #3945 on 1287	1287 mdl 3 or 4 Only ANSCS OCR Size A (10)(11)	
IBM 1428 Font†	ANSCS OCR Sizes A&C	NOF††		7B †††	IBM Selectric™ Typewriter††††	IBM 1403 mdls 2, 3, 7, N1 ††††
0	0 (6)	0 0		0	Basic 1287:	0 A N
1	1			1	0 A N :	1 B 0
2	2	1 p		2	1 B 0 ;	2 C P .
3	3			3	2 C P .	3 D Q ⌐
4	4	2 d		4	3 D Q ⌐	4 E R /
5	5			5	4 E R /	5 F S –
6	6	3 ⩊		6	5 F S –	6 G T *
7	7			7	6 G T *	7 H U $
8	8	4 H		8	7 H U $	8 I V &
9	9			9	8 I V &	9 J W
C (4)	C (5)	5 ⩊0		blank (3)	9 J W |(7)	4 K X
N	N				4 K X –(8)	⌐ L Y
S	S	6 blank			♪ L Y ⅀(9)	⌐ M Z
T	T				⌐ M Z blank	blank
X	X	7				
Z	Z				With #3850:	
│ (VFM)(1)	│ (LVM)(1)	8			+ ⅄	
♪	♪				= ⌐	
	⩊	9			{ ·	
⤹ (2)	⩊				} ″	
blank (3)	blank (3)					

†Digits 0-9 only available in 1428 E (Elongated) for imprinting.
††National Optical Font shown by permission of National Cash Register Co.
†††7B shown by permission of Farrington Manufacturing Co.
††††or equivalent.

(1) Recognized and transmitted in document mode only (LVM-Long Vertical Mark, VFM-Vertical Field Mark).
(2) Recognized and transmitted in tape mode only (1428 font).
(3) No blanks are transmitted in ANSCS OCR Size C, Farrington 7B, or 1428 E fonts.
(4) In 1428 font the characters C, N, S, T, X and Z may only occupy the units position of the field or line.
(5) In ANSCS OCR Size A font, the characters C, N, S, T, X and Z may only occupy the units position of the lines in journal mode (mdls 2 and 4).
(6) Only the characters 0-9 are available in ANSCS OCR Size C font.
(7) For machine printed fields a pre-printed long vertical mark is permitted.
(8) Group Erase permits ignoring a line or field. Symbol must be at least .300" long over-striking at least one standard height character. It must overstrike the first character on that end of the line on which scanning is initiated.
(9) Character erase not in the ANSCS OCR Size A character set. Available to ignore a character and its space for typewriters.
(10) Two reading modes are available, numeric defined and alphameric intermixed.
    The following characters are available in numeric defined mode:
        0 1 2 3 4 5 6 7 8 9 . , / – ♪ * ⩊ ♪ ⌐ │ (LVM) ⅀ and with ≈3850 + = ⅄
    The following characters are available in A/N Intermixed mode:
      All numeric defined characters except ⩊. In addition, A-Z, & : ;
      ⌐ and with #3850, + = { } ⅄ ⌐ ' ″

data to the channel, it cannot be interrupted without risk of losing data. For efficient operation the reader should be assigned high priority on the multiplexor I/O channel to which it is connected. Under the best operating conditions, the 1287 can read any of the following:

1. A cash register tape 3 inches wide, with four 10-character lines per inch, at approximately 2250 lines per minute
2. Documents 2.25 x 3 inches, with 30 machine-printed characters grouped in four fields, at approximately 485 documents per minute
3. Documents 6 inches long, with 55 handwritten characters grouped in 28 fields, at approximately 170 documents per minute

The reading station is a lighttight chamber in which the document (or tape) is stopped and registered. A CRT scans the document with a light beam. The document reflects the light into sensors that carry the scanned image to the character recognition circuitry of the 1287. If the 1287 cannot identify a smudged, broken, or poorly formed character after the first recognition scan, it automatically initiates up to ten rescans of the character. If the rescanning is unsuccessful, the program or the operator can select either a halt (for on-line correction by the operator) or automatic marking of the unreadable character and an advance to the next character.

Characteristics of all 1287 models follow:

1. Multiline, two-direction reading of numeric 1428E or ANSI OCR size A type fonts produced by high-speed printers, IBM Selectric typewriters, or from preprinted forms
2. Formatting under processor program control
3. Automatic rescan of unreadable characters
4. Character display on CRT with selective on-line correction of unreadable characters from the operator's keyboard
5. Document counter

Models 1 and 2 can read the preprinted vertical field mark (long vertical mark), and

1. The digits 0 through 9 plus the alphabetic characters C, N, S, T, X, and Z printed in 1428E font,
2. The digits 0 through 9 plus the three abstract symbols "hook," "fork," and chair" printed in ANSI OCR size A font.

Models 3 and 4 can read the preceding characters and also the alphanumeric ANSI OCR character set.

Up to eight 1287 optical readers can be attached to a multiplexor or selector channel of a System/360 Model 25, 30, 40, or 50 processor. The 1287 contains its own controller; a control-unit position on the channel is required. The channel required on the Model 25 is a special feature. Model 30 must have at least 16,384 bytes of main processor storage.

The numeric handwriting recognition available to all models enables the 1287 to read numeric characters 0 through 9 and special symbols C, N, S, T, Z, and X when these have been properly handwritten on cutform documents. Character shapes and spacing must conform to certain basic rules, as outlined in Table 5-1. This feature also enables the 1287 to read the 3/16-inch Gothic font often used for serial numbers on sales checks, etc. Any Gothic font intended for reading by the 1287 must be approved by IBM; a sample of such a font (3/16-inch) is shown here:

0123456789

The 1287 cannot read the hand-printed letter N; and the letters C, S, T, and Z, when used, must be in the units position of the field. The X can be anywhere.

CONTROL DATA, UNIVAC, HONEYWELL, AND OTHERS

Control Data Corporation was also one of the pioneers in optical character recognition; Jacob Rabinow designed a reader about the same time that David Shepherd did, and associated himself with Control Data for manfacturers of OCR devices. In the past, Control Data has been known for the highly successful Model 915 page reader. In 1969, however, after experimentation with a few other models, Control Data delivered the 936-1 document reader, a high-speed, off-line device capable of reading documents at about 1500 per minute (750 characters per second). The scanner utilizes parallel photocells; recognition is accomplished by means of a matrix-matching technique. The single font to be read can be selected from the ANSI OCR-A or OCR-C; the IBM 1428, 1428C, or 407-1; or the Farrington 12F or 7B. Only numeric plus a few special characters are recognized. Mark-sense and/or hand-printed numerics are optionally available; the standard output is magnetic tape. The Control Data reader is one of the fastest and the cheapest of the off-line systems.

The UNIVAC 2703 optical document reading system illustrated in Figure 5-3 reads one line of data from documents or cards printed in the numeric subset of the ANSI OCR size A font or the UNIVAC H-14 numeric font, translates the data to EBCDIC or ASCII codes, and transfers the data via a multiplexer channel to a UNIVAC 9000 Series computer for output to a variety of devices. Mark-read data and punched cards can also be read. The OCR printing, marks, and punched holes can be placed on the same document if certain format considerations are observed.

Fig. 5-3. The UNIVAC 2703 document reader with card reader, paper-tape reader/punch, teletype punch, and data communications terminal.

The UNIVAC 2703 utilizes an array of solid-state photocells that can extract 600 bits or more of high-resolution information from each printed character. The character reading rate is 1500 characters per second if the documents are printed at 10 characters per inch. Six-inch long documents are normally read at a rate of 300 documents per minute. A speed upgrade feature allows the reading of 6-inch long documents at a rate of 600 per minute. It should be noted that the 9000 Series computers have the fastest memory-cycle times for their price class, a distinction attributable to their plated-wire main memories and monolithic integrated circuits. All the foregoing characteristics combine to provide the 2703 reader with a superior throughput capability. The UNIVAC 2703 was first delivered in June 1970.

The Honeywell 243 optical document reader delivered at the end of 1970 reads ten numeric and three special characters from documents or cards and converts the data into electric impulses for input to a Honeywell Series 200 central processor. The 243 reader is designed particularly to handle "turnaround documents" for such applications as billing, subscriptions, and inventory. Documents to be read may be from 3 to 4 inches wide and 3.5 to 8 inches long. After being loaded into an input hopper, they are transported, continuously or on demand at 100 inches per second, past the optical system and into one of three output stackers. Operating control is on-line from a computer. One line of up to 70 characters is read on each document; printed characters must be of the ANSI OCR-A font style.

Table 5-2. Document Readers, 1968–1971.

CHARACTERISTIC	IDENTITY AND TYPE — RECOGNITION EQUIPMENT ERCR WITH INPUT 2; DOCUMENT READER (document reader)	CONTROL DATA 936-1; DOCUMENT READER (document reader)	UNIVAC 2703 DOCUMENT READING SYSTEM (document reader)	IBM 1287 OPTICAL READER (document reader)
Document Handling				
Doc size, inches (width x length)	3.25x3.25- 4.75x8.75	2.25x3.0- 8.5x5.5	2.75x3.00- 4.25x8.75	2.25x3.0- 5.91x9.00
Max char/line	99	80	80	85
Max lines/inch	6	4	1	6
Max lines/pass	2	1	1	Limited only by document size
No. of doc/min	600	1500	300; 600	3200 lines/min
Transport type	Conveyor belt	Friction; vacuum belt	Belt; carousel	Conveyor belt
Feed mechanism	Vacuum	Vacuum	Friction	Vacuum
Sorting facilities	3 output stackers std. up to 9 additional opt.	3 output stackers	3 output stackers	Multistacker (1-13, depending on model)
Recognition				
Scanning technique	Photocell matrix	Parallel photocells	Parallel photocells	Flying spot
Recognition method	Matrix matching	Matrix matching	Matrix matching	Curve tracing
Font styles read	Multifont (any); hand-print std. feature	ANSI OCR-A, OCR-C; IBM 1428, 1428C, 407-1; Farrington 12F, 7B	ANSI OCR-A; UNIVAC H-14	IBM 1428; Farrington 7B; ANSI OCR-A; 3116 Gothic; NCR NOF
Character set	Numeric, alphanumeric (40 recognition masks); numeric hand print	Numeric	10 numeric plus special symbols	Alphanumeric

Table 5-2. Document Readers 1968–1971 (Continued).

HONEYWELL 243 DOCUMENT READER (document reader)	OCR SYSTEMS 1000 DOCUMENTS READER (document reader)	OPTICAL SCANNING CORPORATION, OPSCAN 288 CHARACTER READER (document reader)	ALLIED COMPUTER SYSTEMS MARK I, MARK II (document reader)	DATA RECOGNITION DRC 700 OPTICAL SCANNER ENCODER (document reader)
3.0x3.5-4.0x8.0	3.25x3.5-8.5x11.0	3.5x2.5-8.5x4.5	2.5x3.5-3.5x7.5	3.25x7.375 or less (51-, 80-col card size)
70	80	80 (mach. print); 20 (hand print)	31	14
1 line/doc	4	3 (mach. print); 2 (hand print)	1	1
1	1, 2, or 3	3	1	1
600-1,100	420	200–600	Oper. dpndnt	67
Vacuum drum	Vacuum belt	Drive belt	Friction roller	Pneumatic; friction
Friction	Friction	Friction; vacuum	Manual	Vacuum
3 output stackers	3 output stackers	Dual output stackers	No	N/A
Photocells	Solid-state array	Photocell	Photocell matrix	Not available at publication
Stroke analysis	–	Matrix matching	Matrix matching	Not available at publication
ANSI OCR-A	ANSI OCR-A; ISO OCR-B; Farrington 7B; hand print	ANSI OCR-A; IBM 1403; 1428, 1428E, 407E, hand print numeric plus C, N, S, T, X 2; + (plus); − (minus)	IBM 1428; ANSI OCR-A	Farrington 7B-1
Numeric plus 4 special symbols	Alphanumeric (hand print numeric only)		Numeric	Numeric

Table 5-2. Document Readers 1968–1971 (Continued).

CHARACTERISTIC	RECOGNITION EQUIPMENT ERCR WITH INPUT 2; DOCUMENT READER (document reader)	CONTROL DATA 936-1; DOCUMENT READER (document reader)	UNIVAC 2703 DOCUMENT READING SYSTEM (document reader)	IBM 1287 OPTICAL READER (document reader)
Multifont flexibility	Limited only by 120-char total, font recognition masks per system	N/A (1 font)*	N/A	Mixture of fonts, hand print, mark sense possible
Max reading speed, char/sec	2,400	750	1500	2000
Operating Control, Flexibility	Off-line; self-contained software allows reformatting, reading of selective fields	Off-line; integral CDC processor; field selection by operator; format under computer program control	UNIVAC 9000 Series; limited field selectivity under computer program control	IBM System/360; selective fields, format control under program control
Standard Output Specifications	Magnetic tape	Magnetic tape	Devices compatible with on-line computer	Devices compatible with on-line computer
Auxiliary I/O Devices, Capabilities	Optional line printers, endorser; additional tape units, microfilmer; mark sense std.	N/A	Devices compatible with on-line computer	Devices compatible with on-line computer

*N/A — Not Applicable

Table 5-2. Document Readers 1968–1971 (Continued).

HONEYWELL 243 DOCUMENT READER (document reader)	OCR SYSTEMS 1000 DOCUMENTS READER (document reader)	OPTICAL SCANNING CORPORATION, OPSCAN 288 CHARACTER READER (document reader)	ALLIED COMPUTER SYSTEMS MARK I, MARK II (document reader)	DATA RECOGNITION DRC 700 OPTICAL SCANNER ENCODER (document reader)
N/A	1 font per pass; intermixed hand print accepted	Reads optical fonts plus hand printed character on a single line	N/A	N/A
800	800	800	20	Not significant; printer speed governs thruput
Honeywell 200 Series, except Model 8200	Off-line; Varian Data 6201; 4K memory controller, demand feed	Off-line; plug-board controls reading of selected fields, intermixed	Mark I—integral controller; Mark II—PDP-8L disk system	Data recognition integral processor (hard-wired); rescan feature allows searching for line in several locations
On-line to Honeywell Series 200 computer	7-, 9-channel mag tape	7-, 9-track mag tape (556, 800 bpi)	Mark 1—paper tape, mag tape cards; Mark II any PDP-8L output	Input invoices encoded with E-13B MICR as output
N/A	On-line configurations, media possible; mark-sense capability std; other fonts opt.	N/A	N/A	N/A

Table 5-2. Document Readers 1968–1971 (Continued).

IDENTITY AND TYPE / CHARACTERISTIC	RECOGNITION EQUIPMENT ERCR WITH INPUT 2; DOCUMENT READER (document reader)	CONTROL DATA 936-1; DOCUMENT READER (document reader)	UNIVAC 2703 DOCUMENT READING SYSTEM (document reader)	IBM 1287 OPTICAL READER (document reader)
Error Checks, Safeguards	Programmable actions; rescan features; sorts documents into error stacker	Program-selected actions	Error character substituted; documents sorted to reject stacker	Rescan feature; character display with manual insertion; document marked
Special Features, Comments	Hand-print set includes digits; C, S, T, X; + (plus); – (minus)	Mark-sense and/or hand-printed numerics	Mark sense; punched holes read	Hand-print; mark sense; serial numbering
First Delivery	January 1970	December 1969	June 1970	1968
Pricing				
Monthly rental (including maintenance)	$14,050-19,210 or more (includes line printer)	$1,799-2,114	$1,050-1,600	$3,515-7,264
Purchase price	$550,000-767,375 or more	$105,180-119,970	$42,000-64,560	$122,320-266,550

Table 5-2. Document Readers 1968–1971 (Continued).

HONEYWELL 243 DOCUMENT READER (document reader)	OCR SYSTEMS 1000 DOCUMENTS READER (document reader)	OPTICAL SCANNING CORPORATION, OPSCAN 288 CHARACTER READER (document reader)	ALLIED COMPUTER SYSTEMS MARK I, MARK II (document reader)	DATA RECOGNITION DRC 700 OPTICAL SCANNER ENCODER (document reader)
Programmable actions for unreadable character; double document, read error	Rejects sorted, keyed on-line; check digits; batch balance; programmed checks	Error character substituted for unreadable character; error sorting; no re-scan feature	Check-digit verification parity on some output devices; match line halts for keyed ID of nonread character	Detected, rejected character inhibit; output MICR coding
Mark sense; contiguous, demand feed	Software-controlled recognition process; additional memory possible	Handles stock from 20-100 pounds	Audit trail (batch, sequence numbers)	Dedicated only to reading, printing bank-credit card invoice
November 1970	1970	July 1968	1970	1970
$1,700-2,020	$1,830 plus output device	$1,988-2,836	Variable, based on configuration, number of doc read; $600 minimum	N/A
$67,200-79,800	$69,000 plus output device	$98,088-137,738	N/A (leasing only)	$80,000

65

An optional mark-sense reader is available to read up to 12-level mark-sense marks in addition to the character reader.

The OpScan 288 illustrated in Figure 5-4 optically reads one line of alphanumeric data from documents printed in the ANSI OCR-A font or hand-printed characters, subject to certain restrictions similar to those that apply to hand printing for the IBM 1287 optical reader. A single row of photocells is used to scan a single line of print from documents "on the

Fig. 5-4. The OpScan 288 document reader.

fly." Reflected light from the scanner is translated into characters by matrix-matching logic circuitry. The printed line may contain up to 40 hand-printed characters, up to 80 machine-printed characters, or a combination. Optical Scanning Corporation (OpScan; Digitek) had considerable experience with mark readers before announcing a character-reading device.

Four recently delivered systems from smaller, recently established companies are compared with the older companies' systems in Table 5-2. Each system is notable in one way or another. The OCR Systems, Inc. device provides hand-print and software-controlled recognition; Allied Computer Systems produces a manually fed device (available only by leasing), while Data Recognition produces a device exclusively designed for printing bank-credit card invoices, which can only be sold; and the Orbital Systems unit is actually a marriage of a reader supplied by Orbital Systems with a magnetic-tape data record made by Sangamo Electric Company, Transitel Division.

6. PAGE READERS

The recent proliferation of limited capability document readers can be expected to continue for the next few years. However, developments in the field of page readers, which tend more and more toward multifont and omnifont capabilities, may very well outstrip the document readers in the final analysis, since machine prices are going down and the costs of multifont recognition and multipurpose transports are also going down. Nevertheless, the distinction between the page and document readers may linger a long time because both prices and technology have a long way to go.

The older manufacturers of page readers are understandably enough the same manufacturers who initially developed document readers: Farrington, Recognition Equipment (REI), and IBM. There are two notable exceptions: The Control Data 915 page reader, delivered in 1964, is the offspring of research in the 1950s and did not result in a concurrent document reader until recently; and Scan Data Corporation, a relatively new and exclusively OCR company that has made sufficient inroads on the market to be classified as a major OCR manufacturer. Current trends in page reading are exactly opposite to those of document reading: The direction is toward more versatility in font recognition. The sudden proliferation of devices utilizing combination page-and-document transports is such that Chapter 7 will be devoted to them; here the older devices and the increase in font capabilities will be discussed.

FARRINGTON: 3000 SERIES PAGE READERS

The Farrington 3000 Series page readers Models 3030 and 3050 read multiple lines of data printed in OCR fonts on printed pages and output the data in the form of punched cards, punched paper tape, or magnetic tape. Only magnetic-tape output is available from the 3050 reader. The input to both models can be source data produced by a typewriter or other printing device, and/or turnaround (reentry) documents produced by a

Table 6-1. Farrington Document and Page Reader Characteristics.

INPUT/OUTPUT SPECIFICATIONS	3050 PAGE READER	3030 PAGE READER	3010 DOCUMENT READER
Input			
Standard Input	Printed pages	Printed pages	Printed documents or cards
Length (inches)	5.60-14.00	5.60-14.00	2.00-6.00
Width (inches)	4.50-8.50	4.50-8.50	2.25-8.50
Thickness (inches)	0.004-0.058	0.004-0.058	0.004-0.010
Weight (pounds)	24-28	24-28	24 minimum
Fonts: capabilities	Farrington 12L; ANSI OCR, size A	Farrington 12L; ANSI OCR, size A	Farrington 12F, 12L, and 7B; IBM 1428 ANSI OCR, size A
Max. no., programmed	One	Two	Five
Max no. on page	One	Two	One
Auxiliary input	Keyboard	Keyboard, magnetic tape, punched paper tape, punched cards	None
Recognition logic	Hardwired	Hardwired	Hardwired
Control logic	Program plugboard	Computer	Program plugboard
Basic orientation	Alphanumeric	Alphanumeric	Numeric
Output			
Magnetic tape 7-track	Yes; 200, 556 and 800 bpi	Yes; 200, 556 and 800 bpi	Yes; 200, 556 and 800 bpi
9-track	Yes; 800 bpi	Yes; 800 bpi	Yes; 800 bpi
Punched cards	No	Yes; 80-column	Yes; 80-column
Punched paper tape	No	Yes; 5, 7, 8 channels	No
Listing accumulator tape	No	No	Yes; 8-column

high-speed printer operated on line with a computer. The familiar, more versatile and elaborate 3030 page reader and the new, lower-priced, single-unit 3050 page reader are similar in many respects; in particular, both employ the stroke-analysis recognition technique characteristic of Farrington readers. See the discussion on the Farrington 3010 document reader in Chapter 5. For the basic differences and similarities among all three Farrington readers, see Table 6-1. The 3030 and 3050 readers are illustrated in Figures 6-1 and 6-2, respectively.

Fig. 6-1. Farrington 3030 reader.

In 1955 Farrington first marketed an optical reader that read documents printed in one of several fonts and produced 180 cards per minute. Its first page reader was installed in 1959. The current 3030 reader can discriminate two fonts; it utilizes a stored-program computer with a memory capacity of 4,096 to 32,768 positions and an extensive instruction repertoire to implement considerable editing and processing control. It also provides a CRT display and a keyboard for the identification and correction of questionable characters.

The same two fonts are compatible with both readers, namely, the Farrington 12L and the American Standard optical font (ANSI OCR), size A. As selected by the user, the 3050 reader is wired for either of these fonts, but the hard-wired recognition logic of the 3030 can be programmed to read one or both. In addition, the stored-program computer of the 3030 reader confers various editing and processing capabilities, as previously mentioned. Programs can be read into the computer from magnetic tape, from typed pages, and from punched paper tape. Input data can be entered from punched cards, punched paper tape, magnetic tape, and at

Fig. 6-2. Farrington 3050 reader.

the keyboard as well as from the intended medium of printed pages. The outupt device, in turn, can be a seven- or nine-track magnetic-tape drive, a card punch, or a paper-tape punch. This reader can also handle pages on which the two fonts are mixed, with a minimal rejection rate usually prevailing when there is programming to specify the font occupying a particular field.

The 3050 page reader accomplishes the basic reading and error correction functions of the 3030 reader, but its hard-wired recognition logic is limited to only one or the other of the two possible fonts. In addition, many features and auxiliary output devices of the 3030 reader have been withheld from the 3050, consistent with its economy objectives. Nevertheless, this reader is a full-fledged optical character reader providing magnetic-tape output.

The character-reading rate of both readers is stipulated as 400 character spaces per second and the rejection rate as 0.1 percent or less per line. To realize these standards, the machines require careful preparation of the input pages with respect to paper quality and thickness, print definition, print line regularity, and evenness of print and line spacing. In accordance with these conditions and the page size, the readers should attain throughputs ranging from 5,000 to 25,000 lines per hour.

RECOGNITION EQUIPMENT, INC.: INPUT 80 PAGE PROCESSOR

With the announcement of the Input 80 page processor, Recognition Equipment, Inc., reaffirmed its original marketing emphasis on high-speed, high-performance, high-priced page readers. The Input 80 system is an even faster and more flexible multifont reader than the ERCR (Electronic Retina® computing reader). Ironically enough, it handles only pages, even though interest in composite systems is now at a new high, whereas the older ERCR could handle both documents and pages at a time when no other composite system existed.

The Recognition Equipment Input 80 page processor, delivered in 1971, is a stand-alone, high-speed, modular page-processing system with a newly designed Integrated Retina® and paper-handling system. Multifont data obtained from pages with carbon backing, staples, paper clips, dog-ears, folds, and pasted-on labels are read at speeds up to 3600 characters per second and are written on seven- or nine-track magnetic tape after being processed. A 600- or 1000-line per minute printer can also be included in the system. The standard 16K controller can be expanded up to 32K in increments of 8K; an optional page-sequence numbering device allows printing of batch and sequence numbers on pages as they are read. Recognition of an extended numeric hand-printed character set and mark-sense pencil marks is an optional feature on all models. There are three standard systems: a "single font" system, recognizing only one font from the list in Table 6-2; a "multiple font" system that can recognize up to nine fonts chosen from the list in Table 6-2, but which must be programmed to expect a particular font in a given location on a given page; and finally, a "multifont" system that will recognize any fixed-pitch font with appropriate printing specifications and that needs no prior programming. The single- and multiple-font standard systems are equipped with a vocabulary of 40 basic character reference patterns, which may be 10 characters each in four fonts, 40 characters in one font, or any combination the user desires; the recognition vocabulary may be expanded to 360 reference patterns in any combination in increments of 10. The multifont standard system is equipped with a basic vocabulary of 120 character reference patterns, expandable in increments of any 10 patterns up to 360 character reference patterns.

Table 6-2. Characteristics of Font Styles for Input 80 Single and Multiple Font Systems.

FONT STYLE	NO. OF CHARACTERS	PITCH	CHARACTER TYPES
ANSI OCR-A, size A alphanumeric	51	10	Alpha, numeric, special
ANSI OCR-A, size A numeric subset	10	10	Numeric
ISO OCR-B size I alphanumeric	56	10	Alpha, numeric, special
ISO OCR-B, size I numeric subset	10	10	Numeric
IBM 1403 Standard alphanumeric	51	10	Alpha, numeric, special
IBM 1403 Standard numeric subset	14	10	Numeric, special
IBM 1403 Modified (X03)	51	10	Alpha, numeric, special
IBM 1428 OCR	44	10	Alpha, numeric, special
IBM 1428 E	19	7	Alpha, numeric, special
Farrington 7B	18	7	Alpha, numeric, special
NCR 407E-1	12	9	Numeric, special
ABA E13B (MICR)	14	8	Numeric, special
Anelex	48	10	Alpha, numeric, special
3/16 Gothic	10	7	Numeric

The new Integrated Retina® deserves mention, since it differs from the photocell matrix used in the Electronic Retina® employed in REI's previous equipment offerings. This new retina employs a high-intensity light beam and two synchronized mirrors to project 36 slices of each character onto a "retina" of 96 photodiodes incorporated into a slice of silicon. The resultant decision points are defined in terms of 16 shades of gray and are compared with 24 adjacent points to adjust character images and sharpen contrast. Each adjusted character is then compared to pre-stored patterns in memory for proper identification. Recognition of hand-print characters, on the other hand, is based on a feature analysis of curves, vertical and horizontal lines, intersections, and so on. Adjustments for size variations, line skew, and vertical misregistration of characters are automatically made by the Integrated Retina® for both hand-printed and machine-printed characters. The precision of the paper handler is an important element of the high speed, flexibility, and low reject rate expected of the Input 80 system; a combination of vacuums, air jets, belts, and rollers is employed to cut down relative movement between paper and belts, thus minimizing problems (such as buckling) found with most friction-belt feeders.

IBM 1288 OPTICAL PAGE READER

The 1288 is IBM's first venture into page reading; previous OCR devices have been document readers, journal-tape readers, and card reader/punches. It is similar to the 1287 in several respects: The standard input font is the full alphanumeric ANSI OCR-A rather than one of IBM's own OCR fonts, and all optional features (including numeric hand print) available to the 1287 are available to the 1288.

The 1288 reads a variety of document sizes in addition to standard 8.5 x 11 inch pages; documents may range in size from 3 x 6.5 inches to 9 x 14 inches (legal size). Documents are fed under program control from the input stacker, which can hold a stack of documents up to 10 inches high. The 1288 separates the documents, checks for a double-document condition, aligns the documents, reads the document with a flying-spot optical scanner in one to four blocks (called "read increments"), and transports the documents to one of two output stackers, each capable of holding a 4.5-inch stack of documents. As is true with other IBM document readers, appropriate manuals should be consulted when choosing paper and printing ink for the documents.

Two modes of use are possible: formatted and unformatted. In the formatted mode, program control allows the reading of variable-length fields in any sequence. The characters in the field can be oriented in a "normal" direction, or rotated 90 degrees. Scanning is from right to left; the scanning beam is directed by the program. In the unformatted mode, multiple and continuous lines of alphanumeric characters up to six lines per inch and right or left justified can be read in the "normal" direction only. The scanning beam automatically seeks the first character and scans the line. No preprinted reference mark, as in the formatted mode, is required.

Since the scanned area "read station window" is only 6 x 8.25 inches, large documents must remain in the read station for a longer period of time while the read station window is moved. The position of the read station window is directed by the program to four discrete locations, called A, B, C, and D. All or any one of these read increments can be programmed.

Table 6-3. Page and Microfilmed-Page Readers, 1968–1971.

IDENTITY AND TYPE / CHARACTERISTIC	RECOGNITION EQUIPMENT INPUT 80 (page reader)	IBM 1288 OPTICAL PAGE READER (page reader)	FARRINGTON ELECTRONICS PAGE READER MODEL 3050 (page reader)	INFORMATION INTERNATIONAL GRAPHICS I (page reader–microfilm) same as COMPUSCAN	OCR SYSTEM COMPUSCAN 370 (page reader–microfilm)
Document Handling					
Doc size, inches (width x length)	4–14x5.34–9	3.0x6.5–9.0x14.0	4.5x5.6–8.5x13.5	Documents microfilmed before input	16, 35 mm sprocketed roll microfilm, negative image
Max Char/Line	– – –	90	75	Varies relative to reduction ratio	Varies relative to reduction ratio
Max lines/inch	6 (uppercase); 3 (lowercase)	6	6	Varies	Ratios up to 45X
Max lines/pass	– – –	84	70	Varies	Varies widely
No. of doc/min	– – –	14–328	150–400 lines/min	Varies	Varies
Transport type	Vacuum; air jets; belts; rollers	Conveyor belt	Drive rollers	Computer film	Computer film transport
Feed mechanism	Vacuum	Friction	Vacuum	N/A	N/A
Sorting facilities	Not available at time of publication	2 output stackers	Dual output stackers	N/A	N/A
Recognition					
Scanning technique	Integrated Retina®	Flying spot	Mechanical disk	Flying spot	Flying spot
Recognition method	Matrix matching	Not available at publication	Stroke analysis	Matrix matching; feature extraction	Matrix matching; feature extraction
Font styles read	Multifont (any); hand print	ANSI OCR-A; hand print	ANSI OCR-A; Farrington 12L	Unlimited	Unlimited

Character set	Alphanumeric, numeric; hand-print numeric	Printed alphanumeric; hand-printed numeric	Alphanumeric	Unlimited	Unlimited
Multifont flexibility	Limited only by total of 360 recognition masks	N/A	N/A	Unlimited	Unlimited
Max reading speed, char/sec	3600	1000	400	Varies	400 (multifont); 1000 (1 font; alpha); 4000 (numeric)
Operating Control, Flexibility	Off-line; 16K-32K processor; multi-font models intermix w/o preprogramming; multiple-font models must be preprogrammed.	On-line to any IBM System/360 computer; formatting, selective fields under computer program control	Off-line; plugboard controls data skipping, reading of selective fields	Off-line; self-contained PDP-10 computer (32–96K); extremely flexible format handling; image processing; any font, any size	Off-line; self-contained Sigma 3 computer; fonts, font size may vary on page; fixed pitch; proportional spacing
Standard Output Specifications	7-, 9-track mag tape (200, 556, 800, 1,600 bpi) BCD, EBCDIC std	Data to computer	Magnetic tape	Any output compatible with PDP-10 computer	Any Sigma 3 computer-compatible device
Auxiliary I/O Devices, Capabilities	Page sequencing unit; line printer; mark sense opt	Mark sense	N/A	Any PDP-10 computer-compatible device	Any Sigma 3 computer-compatible device

75

Table 6-3. Page and Microfilmed-Page Readers, 1968-1971. (cont.)

IDENTITY AND TYPE / CHARACTERISTIC	RECOGNITION EQUIPMENT INPUT 80 (page reader)	IBM 1288 OPTICAL PAGE READER (page reader)	FARRINGTON ELECTRONICS PAGE READER MODEL 3050 (page reader)	INFORMATION INTERNATIONAL GRAPHICS I (page reader—microfilm) same as COMPUSCAN	OCR SYSTEM COMPUSCAN 370 (page reader—microfilm)
Error Checks, Safeguards	Under program control; lines marked; document to reject stacker; error indicator on tape record; etc.	Rescan feature; reject stacker	Character display with manual correction by keyboard; rescan feature; marks documents	Completely programmable; CRT display with manual insertion possible	CRT display with manual insertion; special symbol substituted for unreadable character
Special Features, Comments	3 system types: single font, multiple font (up to 14 fonts), multifont (any font); all models extensive editing and reformatting	Serial numbering; reads unformatted data	———	New fonts "learned" by reading in, identifying via special CRT keyboard; extensive software for discriminating, interpreting new fonts	New fonts can be entered from documents read; reference matrix may be modified by light pen, CRT
First Delivery	August 1971	1970	August 1969	Mid-1971	Late 1970
Pricing Monthly rental (including maintenance)	$11,895–17,220 or more	$5,980–7,550	$2,345	Not available at publication	$18,000
Purchase price	$446,000–665,000	$223,390–296,480	$140,000	$1,500,000	$900,000

Thus, only parts of a document need be scanned if that is desired. As the number of read increments is increased, the throughput of the 1288 drops.

The 1288 can be used on System/360 Model 25 processors equipped with a multiplexor or selector channel, a Model 30 having 16,384 or more bytes of core storage, and Models 40 and 50; up to eight can be attached per system. A high priority should be assigned, since once the 1288 is commanded to read, bytes are transferred synchronously with the 1288 timing.

International Business Machines specifies typical document through-put rates as ranging from 328 documents per minute for a 3.25 x 7.375-inch document of 12 characters on each of two lines, to 14 documents per minute for an 8.5 x 11-inch document of 65 characters on each of 50 lines.

CONTROL DATA 915 PAGE READER

Although it was first delivered as early as 1964, the Model 915 page reader has enjoyed substantial success right up to the present day. Like the Recognition Equipment ERCR, the 915 employed photocells in its scanning technique rather than the mechanical disk used by Farrington

Fig. 6-3.   Control Data 915 page reader.

and IBM at the beginning of the decade; and like the ERCR, the 915 has been able to maintain high speed and sufficient flexibility to remain competitive up until the present. Table 6-3 compares page readers delivered between 1968 and 1971, and it is noteworthy that the Control Data Model 915, which sells for approximately $84,000 plus the cost of the controlling computer and output device, is still less than half the price of the IBM 1288, which reads the same font (at, however, three times the speed). For a page reader in the RETYPE mode, the Control Data Model 915 may still be a good buy for many applications.

The Control Data 915 optical character recognition (OCR) page reader, illustrated in Figure 6-3, is a general-purpose, program-controlled optical reader that reads data from all or part of printed paper documents at a rated speed of up to 370 characters per second. It can record the data on all or any of the following output media: magnetic tape, paper tape, punched cards, and line printer forms. A data communications interface is also available. The input can be original data from a typewriter or other printing device or a turnaround (reentry) document produced by a computer system. Programs stored in the associated computer allow extensive editing and fomat control.

The Model 915 can function as a small, complete data processing system suitable for record processing and storage or as a processing terminal associated with a larger computer system. Basic differences between the three processors available with the 915 are as follows:

1. The 8092 is a small economical controller, which primarily controls the 915.
2. The 8090 and 1704 computers add high-speed computing capability to the 915.

Configurations for the 915 may vary considerably, depending on the capabilities of the processor and the needs of the user. Of the three typical configurations, the 915/8092 system might be called the basic one, since it combines limited computing capabilities with either magnetic tape or communications output. The input documents may vary from 4.25 to 12 inches in width and from 2.75 to 15 inches in length, but the document size may not vary within a batch. Documents should be longer than they are wide unless the exceptions are approved by Control Data. Continuous forms from 4.25 to 12 inches in width can also be read. The paper must be from 0.003 to 0.007 inch thick and from 14- to 38-pounds substance weight, with an opacity of 80 to 97.

Lines of data are read sequentially by the 915. Entire lines or parts of lines may be skipped under program control, and also as directed by filled and unfilled circle indicators. The lines of data to be read may be spaced 6, 5-1/3, 5, 4, or 3 per inch. A minimum 0.38-inch top margin must be allowed for 3 lines per inch spacing and a minimum 0.25-inch top margin for 6 lines per inch. The maximum reading length of lines is 11 inches.

The 915 reads data in two basic modes based on these line spacings: 3 lines per inch and 4 to 6 lines per inch. When 3 lines per inch are being read, a cumulative misregistration of 100 percent per line, or 25 percent per character, can be tolerated. The 100 percent misregistration represents the maximum reading height of 0.318 inch, or three times the nominal character height. When the line density is greater than 3 lines per inch, 0.176 inch of the maximum reading height (called a "window") is scanned. In each line, the first character is assumed to be within 0.03 inch of being centered. If it is not, a 0.012-inch window adjustment is made (see Fig. 6-4). Also, the document can be moved 0.004 inch in the reading station for each of the first six characters in a line.

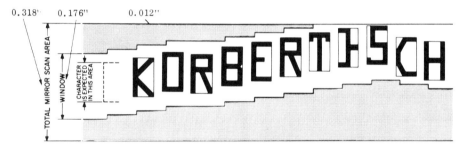

Fig. 6-4. Window-opening adjustments made for a skewed line on the CDC 915.

Documents are moved from the input hopper, through the deskewing and reading stations, and into an output hopper by a series of vacuum-friction belts. The belts are clutch-engaged to two stepping motors and to a takeaway motor. Twelve photocells monitor document progress and check for the following conditions: empty hopper or full stacker, proper document alignment, proper spacing between documents, and double document feeds. Mechanical fingers straighten the documents at the deskewing station.

Scanning and Recognition

Light reflected by the documents and passed through the six light pipes to the right is quantized into on-off impulses to be stored in shift registers. The line-scanning mirror reads the characters from left to right, and the high-speed oscillating mirror scans vertically about three times during the time interval required for the scanning mirror to move one stroke width (nominal 0.014 inch) to the right. The contents of the shift register are then matched against register diode matrices representing the character-reading set.

The character set is the standard set for the ANSI OCR-A font, including 26 letters, 10 numbers, and 33 symbols. Lines may be skipped fully or partially under program control. Data may also be mark-sensed by marking or not marking in preprinted circles.

Checkpoints on the input documents are used as signals for the computing unit to start or stop reading and to skip unwanted data. Vertical field separators and filled circles are frequently used as yes/no indicators for this purpose. Special characters may also function in this manner; but they are more frequently used to identify such fields as account number or amount, or to simplify editing routines.

The 8092 Teleprogrammer is a small economical controller providing 4096 eight-bit words of storage. It has an access time of 4.33 microseconds and a 45-instruction repertoire. The 8090 computer provides from 4,096 to 32,768 twelve-bit words with an access time of 6.4 microseconds and a 130-instruction repertoire. The 1704 can also contain 4,096 to 32,768 words of storage. Its 18-bit words contain 16 data bits, 1 parity bit, and 1 program-protect bit. It has a 1.1-microsecond access time and a 72-instruction repertoire.

Data are usually recorded on magnetic tape, but they can also be recorded on all or any of the following output media: paper tape, punched cards, and printed forms. All processing or writing of data is under program control. Reformatting and editing, accumulating and checking totals, check-digit verification, and other preliminary processing can be performed. Eight software support packages are available for various applications and equipment configurations.

OMNIFONT READERS

All devices discussed thus far have been single font, multiple font, or multifont devices that utilize one or more sets of permanently stored reference patterns against which to compare and "recognize" input patterns. Three companies currently market omnifont readers capable of "learning" new fonts in a few minutes: Scan Data Corporation, CompuScan, Inc., and Information International.

The Scan Data readers are multifont page or page-and-document readers with an optional software package that manipulates hard-wired features to configure a new reference mask when a new font is read into reference memory; these are discussed at length in the chapter on page-and-document readers. Both CompuScan and Information International readers share several unique features in system design as well as omnifont capabilities, and both can microfilm input pages in order to combine jam-free input with the ability to manipulate image sizes.

CompuScan and Information International are using their first proto-
types as the basis of service bureaus capable of handling graphic as well as
character input, but these readers are enormously expensive (see Table
6-3). The Information International reader has more extensive software,
greater processing capabilities and a higher price, but as far as basic system
conception goes, the following discussion of the CompuScan 370 system
is also applicable to the Information International Grafix I system.

CompuScan 370

The CompuScan 370 optical character reading system reads hand-
printed data and data that are typed or printed in any font. It processes
these data for ouptut to any devices of the system that are compatible with
the Sigma 3 computer. In the 370 reader system, all input data are first
transcribed to negative images on 16- or 35-mm sprocketed roll film by
standard off-line microfilming devices, and the frame is then scanned. This
technique holds much promise for microfilm technology; for example, the
film produced by most computer output microfilmers (COM) can be read
directly by the 370 system. Thus, data, whether produced by a COM or by
other means, can be stored on film; when needed as computer input, they
can be digitized immediately by the 370 system and transcribed to a computer-compatible input medium. The 370 is both multifont and "omnifont"; that is, it can read more than one font at the same time, and it can
recognize an unlimited number of fonts of both fixed pitch and proportional spacing. Rated input speeds vary, depending on the difficulty of the
material; a speed of 400 characters per second is average for multifont data
printed in variable pitch; a speed of 1000 characters per second is average
from typewritten alphanumeric data; and a top speed of 4000 characters
per second has been realized with numeric data. These speeds refer to data
on microfilm and do not include the time required for transcribing documents to microfilm.

The use of microfilm negatives as the direct-input medium both facilitates recognition (through greater contrast within the images and through
the capability of varying the magnification of filmed texts) and solves a
number of transport problems (jams, double feeding, and delicate document movements during scanning) that result from the nature, size, and
condition of fragile paper-input documents. CompuScan, Inc., says that
the slight added cost of off-line microfilming is more than offset by the
increase in throughput speeds gained by using microfilm transports, which
can hold several thousand pages on one roll of film (fewer loading operations) and can change frames in a fraction of a second (higher transport
speeds) without ever jamming (reduced downtime). Since the microfilm
produced as an intermediate step between paper and, say, magnetic tape

can be conveniently stored in a small place, large data banks may be accumulated without the costs involved in the storage of paper documents. The concept of using microfilm transports for economy and reliability on an OCR device should be applicable to any OCR system. This approach should be of interest to manufacturers who are attempting to reach the smaller users through less sophisticated, more economical OCR systems.

The flexibility of the system in handling any number of fonts and any mixture of fonts on a single page is a noteworthy characteristic. This flexibility results from the extensive use of software in all aspects of recognition, as well as from the use of a flying-spot scanner, a microfilm transport, a sophisticated CRT with a light pen, and a combination of cross-correlation matrix matching and feature extraction in the recognition logic. Since much of the data needed to recognize a character are stored in a reference section of memory and can be written out on tape or disk to make room for the acquisition of data relating to a new font, font capabilities are virtually unlimited. The reference memory is capable of storing matrices for approximately 800 symbols, which may be characters, feature masks, or special symbols. Basically, recognition of a character involves comparison of a scanner-generated matrix to a matrix residing in the reference memory; font acquisition involves transference of the matrix generated by the scanner into an empty position in the reference memory. Thus, new fonts can be acquired in a few minutes directly from a page being read; the operator either enters the identifying code via the Teletype console or, if the character is less than ideal, magnifies the character on the CRT and fills it in with the light pen before identifying it. Both the recognition process and the process of acquiring a font are described in more detail below.

During the scanning process, a number of factors are evaluated before the actual process of recognition takes place. These include
1. Video calibration of the overall density of the microfilm
2. Measurements of the base-line slope, or skew
3. Coordinate rotation to compensate for the skew of the microfilm image, if any
4. Size calibration and adjustment of matrix, since microfilm images may not be exactly identical in size
5. Field search, under program control
6. Character search
7. Character group definition of a character, according to the weight, width, and
8. Position of the character to be recognized

Evaluation of these factors comprises much of the work of recognition. After scanning is completed, the character matrix produced by the flying-spot scanner is compared to the appropriate group of matrices stored in the reference section of memory. Several partial cross-correlation measurements are made between the unknown and the reference matrices in the selected group. These measurements are then combined to arrive at a final evaluation. Reference matrices are compared in the order of their frequency in English. For a positive identification there must be a defined degree of resemblance between the scanner mask and the identifying reference mask in memory, and also a defined degree of difference between the scanner mask and the second best identifying mask. Part of the flexibility of the CompuScan recognition system is that the degrees of resemblance are software-controlled and may be adjusted for different applications. The size (number of points) of the matrices may also be controlled by software to allow magnification of questionable characters.

The cross-correlation technique previously described is supplemented by an additional recognition cycle that may occur between two members of a "confusion" pair (e.g., numeral 0 and letter O) that have very slight distinguishing characteristics. These characters may be subjected to a feature extraction analysis in which the appropriate feature masks for that pair are moved over critical areas of the character to determine whether certain distinguishing features are present or absent. The feature masks are stored in core in the same manner as reference characters and are applied under software control. Additional software routines are provided for touching characters, broken characters, and other common recognition problems. Software for many other recognition problems is in the process of being developed. When all else fails, the character can be displayed on a CRT for identification, or a "confusion" symbol can be included in the output in the corresponding character position.

PAGE READERS AS OPPOSED TO PAGE-AND-DOCUMENT READERS

In all recent attempts to improve page handling, the high cost of delicate transports has been obvious. Readers with lower reading speeds are considerably less expensive than those with high speeds, and the new idea of microfilm feeding is an obvious attempt to circumvent the problems of paper handling. The increased speed of the new page reader made by Recognition Equipment is accompanied by a very high price tag at a time when OCR prices are generally dropping. Chapter 7 presents several recent experiments in transports, including combination page-and-document readers at lower speeds and lower prices, and very low speed automatically or manually fed terminals at prices that allow OCR to be used in time-sharing rather than in batch-processing modes.

7. NEW TYPES FOR THE 70'S: PAGE-AND-DOCUMENT READERS, TERMINALS

As explained in Chapter 6, transport problems were among the most significant of any facing OCR manufacturers, and these problems led to the development of two basic equipment types, the document reader and the page reader. However, it was obvious from the beginning that one dual-purpose transport that could handle both reader types was the one to have if the price was right and if it could really handle documents without jamming. Recognition Equipment, Inc. (REI), recognized the existence of this need when they advertised both page and document transports that could operate simultaneously on the ERCR (Electronic Retina® computing reader). This configuration was too expensive for most users, however.

It was not until 1970 when Scan-Optics, Inc., announced their 20/20 reader that a true page-and-document reader was actively marketed as such. This reader combines what is essentially a page transport, fast enough to process documents at a speed of 500 documents per minute, with a scanner capable of 2000 characters per second. The 20/20 is a multi-font alphanumeric reader that reads most OCR fonts in both upper and lower case (see Table 7-1) with a "vidicon image dissector." The Scan-Optic system is basically an off-line batch-processing system that sells for the low price of around $100,000. Journal type can be read as an option.

Actually, two true page-and-document transports were already on the market before the announcement of the Scan-Optics 20/20 but were not marketed as such. The IBM 1288 reads documents ranging in size from

84

3 x 6.5 inches to 9 x 14 inches. The Cognitronics Corporation terminal used on its remote optical character recognition (ROCR) service bureau reads all input a line at a time, with the top of the page feeding first and thus allowing both single-line documents and type-filled pages to be processed in the same manner. Cognitronics marketed its devices as the first OCR terminal, however, bringing in a second new class of devices that added a new dimension to the on-line or off-line configuration, previously in use. The Cognitronics ROCR terminal can read discrete pages or documents and can also read continuous forms and journal tape with appropriate options. Its other capabilities include (1) recognition of the data at a central recognition facility and (2) either recording the data on 7- or 9-track magnetic tape to be delivered to the user or transmitting the data to a remote 7-track recorder on the user's premises. All common numeric fonts, a numeric hand-print character set, and the ANSI OCR-A alphanumeric character set can be recognized. Input may be from manual or automatic feeds. Unreadable characters are displayed on a CRT at the Cognitronics recognition facility and are immediately identified by keyed entries. The systems headquarters use a Digital Equipment PDP-8 computer for control and Kennedy incremental magnetic tape recorders for recording. Cognitronics Corporation is marketing the service to small- and medium-sized users and expects clients with as few as three keypunches to realize economies, depending on the application and the distance from the service center. Additional savings are realized by using a special "programmed form" that reads in preprinted program codes, instructing the scanner to skip a specified number of fields or lines, to expect hand printing, and so on.

The Cognitronics System 70 is similar to the ROCR system except that the user purchases both terminals and central recognition unit to set up his own data collection system. David Shepherd, who built one of the first two optical character readers, is currently the president of Cognitronics.

Both the Recognition Equipment (REI) Input 3 terminal and the Infoton Challenger share the capacity to read both documents and pages as well as to act as inexpensive terminals transmitting to a central recording facility. Both systems, however, can have recognition logic within the reading unit itself, and can act as low-speed stand-alone systems. The Infoton device utilizes a manual feed, has a full 280-character CRT text display, and can read documents as small as 1.25 x 4 or pages as large as 12 x 14 (legal size)—about 1320 characters per second for single-spaced text. The REI Input 3 reader, on the other hand, can read one-line documents and "short pages" up to 6 x 9 inches; it uses an automatic feed and has the capability of reading hand printing.

The versatility of the transport on the terminals is a logical element of the terminal function; since it is meant to be a device used in a time-sharing

Table 7-1. OCR Page-and-Document

CHARACTERISTIC / IDENTITY AND TYPE	SCAN-OPTICS 20/20 PAGE AND DOCUMENT READER (page reader; document reader)	INFOTON CHALLENGER (page and document reading terminals)	RECOGNITION EQUIPMENT ELECTRONIC RETINA COMPUTING READER (document reader)	(page reader)
Document Handling				
Doc size, inches (width x length)	3.0–9.0x4.5–14.0	1.25–12.00x4.00–14.0	3.25x3.25–8.75x4.75	3.25x3.25–14.0x14.0
Max char/line	80	110	90	150
Max line/inch	6	3 std, 6 opt	8	6
Max lines/pass	76 (6080 characters)	40 std, 80 opt	2	100
No. of doc/min	508 (4.5x3.5; 200 characters)	6 min (12x14) to 20 max (6x4) incl. manual feed time	1200	24
Transport type	Vacuum belt; roller	Step drive	Conveyor belt	Vacuum drum
Feed mechanism	Friction feed	Manual feed	Vacuum	Vacuum
Sorting facilities	2 or more stackers	N/A*	Multistacker	N/A
Recognition				
Scanning technique	Vidicon image dissector	Electromechanical	Photocell matrix	
Recognition method	Feature analysis	Feature correlation	Matrix matching	
Font styles read	ANSI OCR-A std; IBM 1403E, 1428E; Farrington 7B, 12E, 12L; NCR NOF; E-13B; hand print opt	ANSI OCR-A	Multifont; hand print option	
Character set	Alphanumeric; upper and lower case	57 uppercase alpha-numeric std; additional 26 lower case, European char opt.	Alphanumeric; upper and lower case	

*N/A–Not applicable

CONTROL DATA 955 (page and document reader)	SCAN DATA READER MODELS 250 AND 350 (page and document readers)	RECOGNITION EQUIPMENT INPUT 3 (document reader/terminal)	COGNITRONICS REMOTE OPTICAL RECOGNITION SERVICE (page, document, journal tape readers)	COGNITRONICS SYSTEM/70 (page, document, journal tape reader)
4.875–11.125 x 3.25–12.625	3.0–11.0x7.0–14.0 (20 sq. in surface min.)	2.25–6.00x3.75– 9.00 3.5–6.00x4.0–9.00	3.25x3.5–8.5x unlimited	2.00–8.50 x 3.25–14.00
100	120	50 (20 hand print)	77	76 (38 hand print)
3 std., 6 opt.	6	3 (2.5 hand print)	6	6 (3 hand print)
36 std., 72 opt.	Governed by page size	1 (documents); 22 (short pages)	No limit; docu- ments read line by line	Unlimited
300 one-line doc; 15 full-print pg.	80–160 docu- ments	Varies; 60 (1-inch line) doc/min; 7.5 (4 5-inch)	Up to 50	Variable
Vacuum, belt	Vacuum	Friction	Rollers	Friction
Friction wheel	Friction	Automatic	Friction	Manual (automatic opt)
Two output pockets	3 output stackers	2 output pockets	None	None
Parallel photocells Matrix matching	Flying spot Cross-correlation matrix matching	Rotating disc Feature analysis	Laser/photocells Topological	Laser Confidential
ANSI OCR-A upper case std. 1 see comments for options	Up to 5: ANSI OCR-A, ISO OCR-B, pica, IBM 1428, hand print	ANSI OCR-A; ISO OCR-B; IBM 1403, 1428 modified; hand print	Multifont; hand- written	ANSI OCR-A; all numeric fonts; hand print
OCR-A, ORC-C alpha- numeric; all others numeric	Alphanumeric; and lower case hand print; numeric	OCR-A, OCR-B alphanumeric; all others numeric	Numeric	OCR-A alpha- numeric; all other numeric

Table 7-1. OCR Page-and-Document

CHARACTERISTIC \ IDENTITY AND TYPE	SCAN-OPTICS 20/20 PAGE AND DOCUMENT READER (page reader; document reader)	INFOTON CHALLENGER (page and document reading terminals)	RECOGNITION EQUIPMENT ELECTRONIC RETINA COMPUTING READER (document reader) (page reader)
Multifont flexibility	Fonts may be mixed on a page	N/A	Fonts can be mixed on a page, within a batch
Max reading speed, char/sec	2000	3000; 660 double spaced, 1320 single spaced	2400
Operating Control, Flexibility	Off-line; integral Hewlett-Packard 2114 Control Computer	Off-line hard-wired controller std; flexible format control with mini-computer option	Off-line; self-contained software
Standard Output Specifications	7-, 9-track mag tape	8-bit ANSI II parallel asynchronous output to auxiliary device	Any peripheral device using mag tape, punched cards, paper tape
Auxiliary I/O Devices, Capabilities	Line printer opt; journal tape input	7-, 9-track mag tape (556, 800, 1600 bpi); 4096 (16-bit) word mini-computer; EIA-RS-232C communications interface	Bar code reader-sorter with document transport
Error Checks, Safeguards	On-line char insertion; check digits; batch totals; selective verification by retyping critical data on same page	Ambiguous char display with keyboard entry; 1280-char CRT text display	Reader: Rescans; programmable actions; sorts doc. Page: Rescans; programmable actions; marks doc.

CONTROL DATA 955 (page and document reader)	SCAN DATA READER MODELS 250 AND 350 (page and document readers)	RECOGNITION EQUIPMENT INPUT 3 (document reader/terminal)	COGNITRONICS REMOTE OPTICAL RECOGNITION SERVICE (page, document, journal tape readers)	COGNITRONICS SYSTEM/70 (page, document, journal tape reader)
Up to 3 fonts, pre-programmed recognition 750	Fonts may be mixed on page; omnifont software option 800	Fonts may be mixed on line, a page 75 (30 hand print)	Any font; OCR-A preferred 50 char/sec transmission rate	Fonts may be mixed on a line, a document Variable
CDC SC 1700 processor (18K-32K); selective reading; reformatting, etc., by software control	Off-line; self-contained PDP-8; computer system controls formatting, editing	Off-line; on-line to IBM System 3, 360/20 and up	On-line by computer at ROCR center; formatting field selection; other controls by codes on documents sent	Off-line; PDP-8; equivalent controller
Mag tape: 9-track, 800, 1600 bpi; 7-track, 200, 556, 800 bpi	Mag tape; any PDP-8 compatible device; may be operated on-line	7-, 9-track mag tape	Data to ROCR center then to mag tape, computer at customer location	7-, 9-channel mag tape, (556 or 800 bpi, respectively)
Any peripherals compatible with SC 1700 processor; also journal tape opt.; mirror-image recognition opt.	Any PDP-8 computer-compatible device; note page/doc feed available as field-installed option	Telecommunications interface; mark sense capability	N/A	Paper tape, TTY
On-line character correction, marking pen, software-controlled data checks	Rejected char corrected on-line or document sorted; software	Document containing unidentified char sorted	Unreadable char flagged on CRT's at ROCR center, corrected by operators at center	Check-digit verification; batch totals; char display with keyboard entry

89

Table 7-1. OCR Page-and-Document

CHARACTERISTIC	IDENTITY AND TYPE	SCAN-OPTICS 20/20 PAGE AND DOCUMENT READER (page reader; document reader)	INFOTON CHALLENGER (page and document reading terminals)	RECOGNITION EQUIPMENT ELECTRONIC RETINA COMPUTING READER (document (page reader) reader)
Special Features, Comments		Serial numbering; edge marking; additional core in 4K increments	Std 110-char line buffers; CRT has full edit capability	Reads mark sense and bar code; accumulates totals; either, both transports can be included in system
First Delivery		August 1970	First quarter, 1971	1964
1971 Pricing Monthly rental (including maintenance)		$3,100 or more	$1,200, depending on lease length	$13,765–36,000 or more
Purchase price		$100,000 or more	$35,000 without output device	$545,000-875,000 or more

rather than a batch-processing mode, its cost could not have been justified unless it were able to read a variety of sizes of source-data documents. The lowest prices for these relatively flexible terminals are achieved by using slower transports, eliminating output devices, and storing recognition logic—all characteristics in keeping with a time-sharing mode of operation.

The two important page-and-document readers proper that followed the announcement of the Scan-Optics reader—the Control Data and Scan Data systems—are, however, both large and flexible processors at a higher price. The moderately priced Control Data 955 reader, delivered in November 1970, is a multifont reader with recourse to a variety of software packages and optional fonts as well as numeric hand printing; recognition of the ANSI OCR-A is standard, and journal tape can be optionally read into

CONTROL DATA 955 (page and document reader)	SCAN DATA READER MODELS 250 AND 350 (page and document readers)	RECOGNITION EQUIPMENT INPUT 3 (document reader/terminal)	COGNITRONICS REMOTE OPTICAL RECOGNITION SERVICE (page, document, journal tape readers)	COGNITRONICS SYSTEM/70 (page, document, journal tape reader)
Many software pkg; opt. fonts: OCR-A (lc), OCR-C, ISO-B, NOF, E-13B, 7B, 12F, 1403 H07E-1, 1428, 1428-E, hand print	Continuous forms, journal tape options; SWAMI omni-font software allows "learn-ing" of new fonts.	Suitable for office environment; use by clerical personnel; sep-arate doc feed (single-line doc) and/or page feed (multiple lines) on 1 device	Leased service; terminal receiv-ing mag tape unit computer at customer's location; data by phone lines to ROCR center in New York, Chicago, Los Angeles	Small terminals linked to central output device
November 1970	March 1971	January 1971	—	September 1970
$5,498 (basic stand-alone system); more for opt.	$5,500 (250) $7,500 (350)	$950–2,025	$425-825 (mag tape unit, autofeed) plus $300 min activity charge	N/A
$170,830 as above	$220,000 (250) 345,000– 420,000 (350)	$33,000–68,300	N/A	$33,600 plus output device

the system. This reader combines some of the attractive features of the Model 915, with added flexibility as to input medium. Table 7-1 com-pares the main characteristics of both the page-and-document reading stand-alone systems and the terminals.

The Scan Data readers, although first marketed in 1967 as multifont page readers (Models 100, 200, or 300) and then in 1970 as omnifont page-and-document readers (Models 250 and 350), are all essentially variations of the same system; in fact, the page-and-document transport can be field-installed on all page reader models, effectively changing the model num-ber. These reading systems incorporate several of the characteristics pro-jected for the 1970's, and are particularly interesting because of the relatively low cost of their omnifont capability when compared with the costs of CompuScan and Information International readers. A detailed description of the Scan Data readers, and particularly of their scanning and recognition techniques, follows.

SCAN DATA OPTICAL READERS

All Scan Data readers are multifont and hand-print readers employing perforated-belt, vacuum-platen transport mechanisms and a flying-spot scanner to achieve rapid character recognition and high throughput rates. All models have the same basic configuration and have recourse to the same basic auxiliary input/output devices and options. Models 250 and 350 differ from Models 200 and 300, respectively, in that they have a paper-handling mechanism that accommodates both documents and pages instead of pages only.

Model 200/250, the basic reader, reads from one to five fixed-pitch fonts from a standard list. Models 100 and 300/350 are essentially custom-designed readers capable of handling up to 50 fixed-pitch and/or proportionally spaced fonts. Models 100 and 300/350 are designed for graphics as well as business applications and can output to typesetting equipment; the Model 100 is specifically designed to read telephone directory pages. Table 7-2 outlines the differences between the various models, and Table 7-3 lists the fonts available to 200/250 systems. Of these, up to five fixed-pitch fonts can be selected from the list in Table 7-3. Fixed-pitch

Table 7-2. Model Differences, Scan Data Readers.

MODEL NUMBER	100	200	250	300	350
Standard feed	Page	Page[a]	Page-and-document	Page[a]	Page-and-document
No. of fonts (limitations)[b]	None (more than 5)	1-5 standard[c]	1-5 standard[c]	None (more than 5)	None (more than 5)
Applications software	Graphics	Business	Business	Business and graphics	Business and graphics
SWAMI software (omnifont)	Yes	Yes	Yes	Yes	Yes

[a]The page-and-document feed is field installable on Models 200 and 300, effectively changing the model number.

[b]All fonts can be field-installed.

[c]See Table 7-3.

fonts can be handled at 10 or 12 characters per inch and 6 vertical lines per inch. The minimum "leading" requirement of a proportionally spaced font is one-fourth of the character height. All fonts may contain both upper-case and lower-case alphabetics, numerics, punctuation, and special symbols. The capability of reading a hand-printed character set of 10 numeric, four alpha, and two special characters is also optional.

Table 7-3. Standard Fonts for Reader Models 200, 250.[a]

FONT	CASE[a]	CHARACTER SET
ANSI OCR-A	Upper	Alphanumeric
ISO OCR B	Upper	Alphanumeric
ISO OCR B	Lower	Alphanumeric
Pica	Upper	Alphanumeric
Pica	Lower	Alphanumeric
IBM 1403 Selectric	Upper	Alphanumeric
IBM 1403 Selectric	Lower	Alphanumeric
Hand print	Not applicable	Numeric plus C, A, T, X

[a]Maximum number of four upper-case fonts; maximum of two lower-case fonts; maximum of five fonts per machine, counting hand printing.

The transport mechanisms of these readers are perforated belts moving over vacuum platens. Since the pull of the vacuum holds the document in position on the belt, relatively few mechanical devices are needed, and the system operates at low inertia. Hence, the document can be started and stopped quickly. Four 2-inch stackers—one for input and three for output—are employed, each equipped with a manual width control of 7 to 11 inches. Movement of the document is under software control. Paper is moved in increments of 0.005 inch at a rate of 1200 increments (6 inches) per second. A speed belt operating at 12 inches per second is also available as an option.

A variety of paper sizes and qualities can be used. Through appropriate program control, widths of from 6½ to 11 inches and lengths of 8 to 14 inches are accepted on standard 100, 200, and 300 readers. Models 250 and 350 can handle paper sizes ranging from 3 x 7 to 11 x 14 inches, providing the surface is at least 20 square inches. The ability to read journal tapes from 1 5/16 inches to 6 inches wide or fanfold continuous forms 6½ to 11 inches wide (but not both) is optionally available to all readers. Acceptable paper may weigh from 15 to 30 pounds (that is, based on 500 sheets 17 x 22 inches; usually stated as 17 x 22/500), and may range from 0.003 to 0.006 inch in thickness.

The scanning mechanism of the Scan Data systems utilizes a flying-spot scanner that enables character spaces to be processed at the rate of 800 per second. A spot of light is generated on the face of the cathode-ray tube (CRT) and deflected to form a scan pattern. A lens system focuses the scanning light spot on the document, which reflects it onto a phototube. The voltage generated by the phototube is then transferred through a decision network to the recognition system. The CRT of the Scan Data readers employs a new high-speed, high-resolution phosphor, which emits light that filters out the yellow-orange range of the spectrum so that it is nonreadable. Character normalization (adaption of the effective image size of the character feature to the corresponding field in the recognition storage register) is accomplished electronically by adjusting the size of the scan raster on the CRT. Changes in the recognition threshold level are generated in a video processor. The overall resolution power required of the CRT is modified by using gross document movements (discrete jumps) along the length of the documents.

The recognition technique of the Scan Data readers is called "cross correlation." As this term implies, the character being read is divided into many small elements, which are compared or correlated one by one with features contained in recognition storage. Examples of possible elements are the cross bar of an "A" or an "H" and an open space in an "O" or "E." In the Scan Data system the elements are termed "features" because they are deliberately made to overlap, a redundancy that facilitates the detection of small differences.

In the first correlation step of the Scan Data system, over 400 features defining the incoming character field are successively compared with a catalog (set) of characteristic features comprising positive and negative properties, which are hard-wired in recognition storage as resistance diode networks. A "most likely" match is made for each feature, and the combination (subset) of matches so generated is then compared to the corresponding "ideal" combinations (subsets) that define the recognition characters in computer memory. At the conclusion of this second level of correlation, the incoming character is correlated with the ideal combination of features (that is, with the character) having the fewest disparities. If the number of disparities is excessive, the character is treated as "nonreadable" and it appears in context on a CRO screen for manual identification by the operator. Alternatively, the software program can assign a "confusion" code to it. These provisions eliminate any need to withdraw the document. As consideration of this two-level technique will indicate, such rejection by the system is rare, even under relatively adverse conditions (for example, degraded characteristics and poor paper).

Design improvements on any stored character may be implemented by appropriate programming, and new character images (feature combinations) may be added to the system. Depending on the software, the computer can utilize the output of the stored (hard-wired) feature system not only to effectuate character recognition but also to control character normalization and threshold-level changes, which are both implemented electronically by adjusting the scan-pattern size and intensity. The stored features are also used to determine the end of a character space in both horizontally fixed and proportionally spaced printed material.

STANDARD SOFTWARE RECOGNITION FLEXIBILITY

Under software scan control, the scanning beam of a Scan Data reader can find the document character and lock onto it, thus relaxing the need for precision printing. A combination of software and hardware controls compensates for cumulative-printing and data-positioning errors. Software is also provided to permit field definition and selective data extraction. Versatility in handling different kinds of forms as well as reading several fonts on a single page is another attribute obtainable under software control. These software facilities are contained in a library of programs called TEXTSCAN (for handling textual data), FORMSCAN (for handling fielded-form data), and FORMAT (for rearranging, validating, and reformatting output data).

OPTIONAL SWAMI OMNIFONT SOFTWARE

An optional software package with the acronym SWAMI (SoftWare Aided Multifont Input) allows any of the Scan Data models to acquire a new font or add new characters in a few minutes. Usually a page of characters in the new font, typed in a predetermined order, are read into the reader under the control of the SWAMI program, but modifications involving keyboard identifications are also possible. Once a font has been read in, the reader will recognize that font in any of the documents encountered thereafter. The only limitations on the number of fonts in memory at one time is the size of the memory itself.

This package can also be utilized to lower reject rates from hard-wired recognition, particularly with consistently degraded characters, by implementing a software recognition cycle after the failure of hardware recognition, thus cutting down on interruptions for on-line corrections. Since software recognition is slower than hard-wired recognition, it is not intended for high-volume applications. The software approach is ideal, however, for an application with a varied reading requirement, an application in a state of transition, or a low-volume application that does not justify the installation of an additional hard-wired font. The combination of hard-wired features and software control seems to provide omnifont capabilities at a price much more accessible to users than does the completely software-based system used by the other two companies with omnifont readers; but, however, it does not have such an extensive capability for graphics design and manipulation.

8. JOURNAL-TAPE READERS

One class of optical character readers has exhibited practically no changes in system design since the beginning of the decade in spite of the availability of a sizeable market, namely, the journal-tape readers. These devices read the printed listing tapes or "tally rolls" from cash registers or adding machines, and either (1) encode the data on magnetic tape, punched tape, or cards, or (2) read the data directly into the computer.

Most of the transport and recognition problems that plague the page-and-document readers do not exist for the journal-tape reader. Since the input medium is a continuous paper strip, the jamming, double feeding, and difficult handling of discrete pages are avoided. The encoder is basically numeric so that the complex recognition of alphanumeric character sets is circumvented. The nature of the medium precludes undue handling by inexperienced people in the field and, in the case of cash registers, the paper is protected by being enclosed within the register.

Three companies currently market journal-tape readers: IBM, NCR, and Farrington. All three readers are quite fast (ranges from 800 to 2500 lines per minute). The Farrington reader, which is the newest of the three, utilizes a flying-spot scanner. Until the present these readers have had a market in the merchandising field of adequate size to support continuous production of new readers, but the advent of the new point-of-sale registers and data collection devices may present serious competition. Descriptions of the three readers follow.

IBM 1285 OPTICAL READER

The IBM 1285 optical reader can optically read numeric digits and several upper-case alphabetic characters printed in the 1428 type font on continuous rolls of paper (journal tapes) prepared by adding machines, cash registers, and similar devices. The 1285 is available for use with the IBM 1401, 1440, and 1460 as well as the IBM System/360.

The Model 1285 reads the digits 0 to 9 and the upper-case letters C, N, S, T, X, and Z. Up to 25 characters per line can be read from a tape whose width may range from 1 5/16 to 3½ inches and whose length may vary from 3 to 200 feet. Printing color should be black for optimum operation. This model is available for use with System/360 Models 25, 30, 40, or 50. In each case the 1285 must be equipped with the System/360 adapter special feature. Another special feature, a selector channel or a multiplexor channel, is needed on a Model 25 if the 1285 is to be attached.

The reading station of 1285 contains an electronic flying-spot scanner consisting of a cathode-ray tube (CRT), an optical system, and a photosensitive device. Initially, unreadable characters are automatically rescanned, followed by either a halt for on-line correction by the operator or by automatic marking of the unreadable line and an advance to the next line.

Reading rates of the 1285 are in the range of 1500 to 2500 lines per minute. The formula for determining the speed is

$$\text{Lines/minute} = \frac{60{,}000}{1.7W + 1.9S + 47/L + 1.4C \pm 5\%}$$

where:

W = tape width in inches
S = clear space from right edge to first character in inches
L = lines per inch
C = characters per line

The preceding formula is for use with the IBM 1428 type font. If the NCR type font is to be used, the coefficient of C should be changed to 1.55; then 0.5 must be subtracted from the term $1.55C$. Marking of lines that contain unreadable characters does not appreciably reduce the throughput. An optional feature permits reading the NOF type style developed by NCR; reading of this type style and IBM 1428 type style is interchangeable under operator switch control.

NCR 420-2

The National Cash Register Company (NCR) 420-2 optical reader, shown in Figure 8-1, provides users with off-line conversion of data from narrow journal tapes printed by sales registers, adding machines, etc., to magnetic tape, both 7-channel at 200/556 bits per inch or 9-channel at 800 bits per inch (compatible with the industry). Alternatively, data may be converted to paper tape or may be read directly into the computer for immediate processing. In the on-line configuration, the 420-2 requires no intermediate equipment to convert original entry data into computer language.

Fig. 8-1. The NCR 420-2 optical journal reader with a 329 controller and 334 magnetic-tape handler.

Media for input to the Model 420-2 is in the form of journal tapes from sales registers, accounting machines, or adding machines. The tape is read "on the fly" at a maximum 13 inches per second. Up to four lines can be printed per vertical inch and as many as 32 characters can be read from each line for a nominal character-reading speed of 1664 characters per second. Any fixed-width right margin from 0 to 0.25 inch can be read, and any fixed-width left margin can be read using an appropriately wired plugboard. The maximum readable vertical misalignment is 0.025 inch, or 10 percent of the vertical character space or 20 percent of the total character heights. Characters can be printed at a horizontal density of up to 11 per inch. The tape may be from 1.3125 to 3.250 inches wide and 10 to 1560 inches long. An 8-inch header is required on each roll or strip of paper. The 420-2 is an improved version of the earlier NCR 420-1 optical reader, whose rated reading speed was 832 characters per second, or one-half the reading speed of the 420-2.

An individual reader plugboard is normally wired for each of the different types of sales registers and adding machines to accommodate the varying characteristics of the journal tapes they produce.

The three output units available for use with the NCR 420-2 are also used as peripheral units with the NCR 315 computer systems. The Model 334-101 seven-channel magnetic-tape handler with the Model 329-1 controller buffer and the 739 nine-channel magnetic-tape unit are the most commonly used output units for off-line operation. The NCR Class 371-201 paper-tape punch produces fully punched (chad) paper tape. The 420-2 can also operate on-line with the NCR 310, 315, or Century computer systems, or with other computer systems that have been provided with the proper interface. The output character set consists of ten numerals and six symbols.

A removable, wired plugboard controls the output format and allows up to four different output formats per board. Characters within a line may be output in any desired sequence. Unwanted data may be deleted and constant information may be inserted. Characters and fields may be duplicated in each of the output records. Whole lines may also be inhibited from being output. Header information may be manually entered at the beginning of each journal tape by putting a heavy black mark across the splice that joins journal tapes. The black mark causes the reader to stop during input and allows manual insertion of information.

Up to 32 characters per line may be read into the buffer of the 420-2. An additional 32 numerials and symbols (e.g., date and register number) may be emitted. Thus, the maximum length of an output line is 64 characters.

The reader can be set to handle characters that are unreadable after seven automatic rescan attempts in one of three different ways:

1. *Stop on Reject.* Reading is stopped and the unreadable line is displayed for operator correction. A light appears over the unreadable character. The correction is made by depressing the key representing the desired character. If there is more than one unreadable character in the line, the light will move to the second unreadable character after the first one has been corrected. The tape is marked on the reverse side to indicate the correction.

2. *Run and Mark.* The tape is marked on the reverse side, ½ inch ahead of the unreadable line, for later posting and reentry.
3. *Reject-Fill.* A character to be substituted for all unreadable characters can be selected by setting the Reject Character Fill dial.

The effective reading speed of the NCR 420-2 with magnetic-tape output is limited by the length of the journal tapes to be processed, the quality of printing on the journal tape (which determines the number of rescans), the method selected for handling errors, and the amount of header information to be entered. If a low reject rate is anticipated, method 1 for immediate error correction can be used. Method 2 or method 3 allows higher processing speeds. Users state that if reasonable care is exercised in the preparation of journal tapes, a reject rate of less than 0.5 percent of the lines can be attained.

Splicing pieces of journal tape together increases the effective processing speed by allowing longer continuous runs. The computer program that processes the journal-tape data can be written to handle the entries in the original or reversed order. Processing the entries in reverse order eliminates the need to rewind the journal tape.

Journal tape can be loaded into the reader in about 10 seconds, the magnetic-tape unit can be loaded in about 30 seconds, and the paper-tape punch can be loaded in about 1 minute.

FARRINGTON 4040

The Farrington 4040 multifont journal tape reader, which is illustrated in Figure 8-2, reads both numeric and selected special characters from narrow listing tapes and either writes the data on magnetic tape or transmits it directly to a computer. This reader employs a flying-spot scanning system and the capabilities of a stored-program computer to provide the following: high-speed processing at a rate of 2000 characters per second; multifont recognition; operator identification from a CRT of unreadable characters; and flexible editing capabilities such as single-character and whole-line insertion. Data may be prepared on any adding machine, cash register, or accounting machine that generates compatible journal tape, that is, tape on which one of eight acceptable fonts is printed.

The Model 4040 was preceded by the slower, five-font Model 3040 journal-tape reader that utilized a changeable plugboard instead of a stored-program computer to implement its recognition logic and editing functions. Farrington began development of the 4040 soon after the first delivery of the 3040, which the company withdrew from the market shortly after application of the improved reader.

The Farrington journal-tape reader can read data at 2000 characters per second printed at 10 characters per inch. Any of eight fonts can be read, namely, Farrington 7B, Farrington 12F, ANSI optical fonts (ANSI OCR),

Fig. 8-2.   Farrington 4040 multifont journal-tape reader.

sizes A, B, and C, IBM 1428, IBM 1428E, or NCR optical font (NOF). They include the numerals 0 through 9, up to four letters, and four or six symbols. The character sets for each font are listed in Table 8-1, with differing symbols for the same function horizontally aligned. A given 4040 reader can be programmed to handle from one to eight of these fonts, but a mixture of fonts during a single run cannot be read.

Two input keyboards are included with the Model 4040 journal-tape reader. The 10-key adding-machine keyboard is used to enter the identification of "unreadable" characters displayed on the CRT. Although the control-code keys on multifont machines may have more than one symbol designated on the top of the key, each key nevertheless outputs a single code, since different symbols in several character sets have an equivalent meaning. The alphanumeric console keyboard is used to enter alphanumeric leader information, auxiliary alphanumeric data, or programs. The alphanumeric console keyboard also has a punched paper-tape reader.

A small stored-program computer executes editing, formatting, and computational balancing and performs other simple calculations at the time when data are transcribed onto magnetic tape. Fields can be deleted,

Table 8-1. Character Sets, 4040 Reader Compatible Fonts.

12F	7B	OCR-A OCR-B OCR-C	IBM 1428 1428E	NCR NOF	ALT. NOF	
+		⑀	⊬			
0	0	0	0	0	⊔	= 0
through	through	through	through	through		through
8	8	8	8	8	⑁	= 8
9	9	9	9	9	9	
C = 10		C = 10	C = 10	= 10	10	= 10
N = 11		N = 11	N = 11	= 11	Ⴗ	= 11
S		S	S	d		
T		T	T	P		
]		⑁	X	⑁		
⑂	⑂	⑂	Ƶ	⊓		

duplicated, skipped, or rearranged; constant information can be inserted and arithmetic operations can be performed. Many computations usually performed by a large computer can be carried out by the self-contained computer of the system during the transcription of data, thus releasing the main computer of a facility for more complex processing.

Output from the 4040 can be written on magnetic tape or transmitted directly to a computer. Magnetic tape is written at 45 inches per second in 7- or 9-track IBM compatible format. Seven-track units ordinarily record in BCD at two of three densities, namely, 200, 556, or 800 bits per inch, and 9-track units record in EBCDIC at 800 bits per inch. Programs are also available to output data in other codes as required by the user.

At the rated scanning speed of 2000 characters per second and a printing density of 10 characters per inch, the Model 4040 journal-tape reader will read 6000 lines per minute if the line length is 2 inches. The throughput of an operating reader, however, depends on such considerations as the actual line length on the tape, the actual printing density, the print quality, the quality of the paper, and the selected method of treating unreadable characters. Rewinding of tapes is unnecessary, since the reader can be programmed to read the tape in either direction, forward or backward. The tape may also be spliced, provided the printed lines are not brought too closely together (0.200 inch), the thickness at the splice does not exceed 0.010 inch, and neither the angle nor the offset between the two sides of the splice exceeds 0.333 inch. A 6-inch blank leader and trailer and 0.250-inch margins must be maintained.

9. APPLICATIONS, COSTS, AND PERFORMANCE

Are the present ocr devices useful tools that can help to solve data handling problems, or are they merely interesting items of curiosity? Despite the many current limitations, the answer is that ocr definitely deserves systems consideration, as proved by scores of successful ocr installations. As with all EDP equipment, however, there should be a careful analysis of application suitability prior to deciding on the use of ocr equipment, particularly since

1. Most ocr equipment costs are still extremely high.
2. ocr lends itself much more readily to certain applications than to others.
3. Recent developments in other data transcription devices may provide solutions to data-handling problems at significantly lower costs than the cost of ocr.

Paper-work systems used in conjunction with data processing fall into three distinct classes: in-house, turnaround, and field. Since microfilm is currently being used only as an intermediate step between paper and scanner, it is premature to discuss total microfilm information handling systems in connection with ocr.

IN-HOUSE APPLICATIONS

In-house documents are internally created documents prepared expressly for input to the data processing department. Commonly, a keypunch section receives all such documents, creates tabulating cards, and verifies their accuracy. The data are then read into a computer system for

104

processing. If the documents originate solely from typewriters, the key-punch or key-to-tape operators may key directly from the source documents. In many cases, however, the source documents are handwritten; and if the data are critical, the forms may require checking and retyping by persons familiar with the application prior to delivery to the keypunch or key-to-tape section.

Without going into the economics of buffered keypunching or key-to-tape encoding versus optical character recognition, it should be pointed out that unbuffered keypunching rates tend to be slower than typing rates. Therefore the possibility of eliminating or reducing this type of keypunching effort should be investigated. Whether or not the best method would be to use key-to-tape encoders, buffered punches, or an OCR system depends on the type of the users' application as well as its size. See Figure 9-1 for a flow chart of the time-saving elements of the OCR system.

TURNAROUND APPLICATIONS

Turnaround document systems currently represent the most fertile area for the application of OCR techniques. The documents are computer-printed cards or paper forms that are sent to the field and ultimately returned for further processing. Exceptions such as differences in amount owed and amount paid in a billing application can be encoded on mark-sense or hand-print fields (see Fig. 9-2). A high degree of document quality control is inherent in the system, and the degree of reliability may be considerably higher than that in typed input document systems.

Typical applications involving the use of turnaround documents include mass billings, subscription renewals, utility meter reading, and airline ticket processing. A simplified billing system, Figure 9-3, shows the similarity between punched card and OCR document reading systems in this application. This similarity, plus proven suitability of OCR for turnaround applications, makes the punched card versus OCR decision a matter of economic consideration rather than one of technical feasibility.

FIELD APPLICATIONS

As was pointed out in the discussion on forms design in Chapter 4, the cost of a reader for completely uncontrolled field applications is so prohibitive that it is not economically feasible for users other than those who have extremely voluminous processing, such as the state and federal gov-

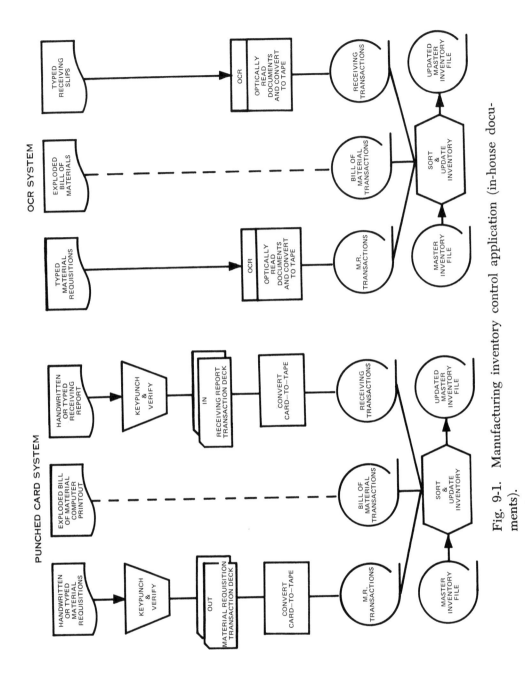

Fig. 9-1. Manufacturing inventory control application (in-house documents).

106

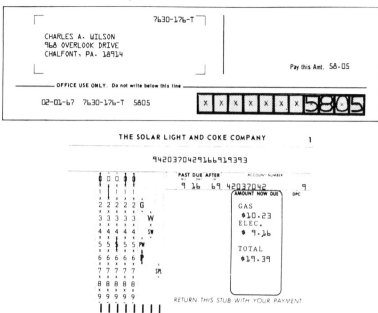

Fig. 9-2. Turnaround documents ready for processing: top, with hand printing; bottom, with mark sense.

ernments. In some instances the careful design of a preprinted form sent into the field may be the difference between success and failure in a field application, even though the document may be encoded by people with a limited knowledge of OCR. As the recognition of hand printing improves, the number of field applications amenable to OCR will undoubtedly increase.

OPERATIONAL PERFORMANCE

A frequent reason why users do not convert to OCR is their lack of confidence in its dependability as an input method. However, user experience indicates that this apprehension is not well founded as long as the OCR system is used within certain constraints. Users in the petroleum industry, the airlines, banking, retailing, utilities, and large service bureaus have said that while OCR is not the ultimate solution to their problems, it adequately performs the desired function—it is the best method currently available for solving their problem.

Reject and error rates of OCR vary with the installation, usually as a function of the application, and the mode of use. Naturally, DIRECT READ applications have higher reject rates. In the RETYPE mode, the degree of proofreading required varies considerably with the need for accuracy.

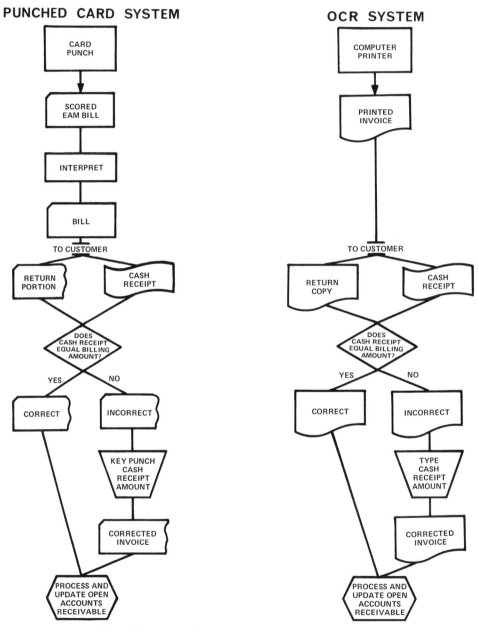

Fig. 9-3. Typical turnaround billing application.

Reject Rates

There is a considerable range in the reject rates that users have experienced with ocr for various types of input data and various degrees of control. When there is good control over the source documents, ocr works well. Controlled conditions exist when input operators are experienced and well trained in the rules for creating ocr documents, and can be directly supervised. Users indicate that they have consistently achieved better than 99 percent acceptance rates for controlled typed input and between 96 and 99 percent acceptance rates for controlled hand-printed numerics.

For those source documents over which some control can be exercised, acceptance rates are typically 90 percent and above. Semicontrolled conditions exist when people preparing the source documents are familiar with the rules of document generation, but who cannot be directly supervised. Examples are large retail outlets where salesmen may fill out point-of-sale slips or branch or remote offices of a large corporation or government agency.

For uncontrolled hand-printed or typed documents, reject rates range from 30 to 70 percent. This situation does not mean that uncontrolled documents can never be profitably read in an ocr system. In certain applications, such as the Social Security Administration system, high reject rates are tolerated, since the portion of documents accepted presents a significant gain over keypunching.

Error Rates

Error or substitution rates indicate the percentages of characters that the ocr machine unknowingly mistakes or misrepresents. Most users believe that their error rates are significantly less than 1 percent of their input characters, but all too frequently they have not monitored this condition. The experience of several large input service bureaus and large commercial users who have run experiments indicates that substitution rates are significantly less than 0.1 percent of the input characters. The Veterans Administration, for example, found an error rate of 0.02 percent on a large test of hand-generated input.

ACCURACY OF PROOFREADING

A fundamental performance consideration in using ocr in the RETYPE mode is the accuracy in the preparation of input documents, by typing and proofreading. Many potential users are unwilling to believe that proof-reading is as accurate as key verification, but user experience again indicates that this apprehension is unfounded.

Users consistently indicate that the degree of accuracy obtainable in typing and proofreading is comparable to that obtained in keypunching and key verification, i.e., 1 to 3 percent error rate. Several large input service bureaus have replaced all their keypunch and key verification units with typing and proofreading, with no complaints from their customers. The fact that many users continue to employ a typing/proofreading sequence demonstrates the viability of this method of preparing input data for computers.

In applying OCR data-handling techniques, however, the impact upon all interfacing systems, equipment, and people should not be treated lightly, for it is a proven fact that the most critical factor affecting the overall performance of today's OCR systems is the condition of the documents being read. This factor will continue to be critical until the development of OCR technology reaches a much higher level.

Under rigidly controlled input document conditions, today's OCR devices can achieve reject rates as low as 2 percent and substitution-error rates of less than 1 percent, both of which are entirely acceptable levels in most applications. In actual practice, however, reject and error rates of about 10 percent and 2 percent, respectively, are much more common in all but turnaround applications. For this reason, a formal systems analysis and forms design effort is recommended to increase potential OCR/computer throughput efficiency, rather than to attempt the retrofit of existing forms or systems. Errors are one of the most expensive problems of computer installations.

Optical character recognition operations are unique in that the document control plays a significantly more important role in reading reliability than does any other single consideration. Various document control techniques that have been applied to OCR to ensure integrity of optically read documents include special data-handling procedures, equipment-operating techniques, and error controls.

ECONOMICS AND SELECTION CRITERIA

In the majority of cases, OCR systems are used as a direct replacement for keypunching operations. Whether or not it pays to make the conversion cannot be answered in any general way, but the following guidelines can be used to assist in the evaluation:

1. What are the characteristics of the specific application—in-house, turnaround, or field-prepared?
2. What are the volume requirements?
3. What present costs are incurred in preparing the data?
4. Can alternative methods be employed?
5. What are the accuracy requirements of the input operation?
6. Is present computer throughput being severely limited by the input media?

A rule of thumb in reaching a preliminary decision on whether to investigate the use of OCR is that an installation preparing 10,000 input documents per day or requiring 10 to 12 keypunch operators is about the smallest that might gain by applying optical reading techniques. As the volume increases beyond this level, ever-increasing savings would tend to result. A useful criterion for making the decision is the number of characters processed per dollar. A simple formula for determining this cost is

$$F = \frac{a}{b + c}$$

where:

$F =$ number of characters processed per dollar
$a =$ total characters processed per month
$b =$ monthly equipment rental and overhead costs
$c =$ monthly employee salary costs, including supervision and
     fringe rates

An example follows for an installation that handles 10,000 documents per day. The parameters are as follows:

No. of keypunch operators $= 12$ Effective hours/day $= 7$
No. of direct supervisors $= 1$ Days/month $= 20$
Operators' salaries (including O/H)   OCR rental $= \$4000$/mo
   $= \$400$/mo   OCR machine O/H $= \$500$/mo
Supervisor's salary (including O/H)   Operator's salary $= \$400$/mo
   $= \$750$/mo   OCR throughput $= 300$ char/sec
Machine overhead $= 10\%$ Characters/document $= 64$
Machine (026) rental $= \$60$/mo   Reject rate $= 10\%$ (rejected
Keypunch throughput $= 7500$ documents must be
   char/hr   keypunched)

The application involves the processing of strictly controlled, field-typed documents.

Case I—Unbuffered Keypunching Operation

$$a = (12)(7500)(20) \qquad\qquad\qquad = 12,800,000 \text{ char/mo}$$
$$b = (12)(66) \qquad\qquad\qquad\qquad\quad = \$\ 792$$
$$c = (12)(400) + (1)(750) = 4800 + 750 = \qquad 5550$$
$$\overline{ \$6342}$$

$$F = \frac{12,800,000}{6342}$$

$$= 2018 \text{ characters/dollar}$$

Case II—OCR Operation

$a = 12,800,000$ (using same volume as in Case I; OCR reader's potential throughput is actually approximately 43 million char/mo)

$$b = 4000 + 500 \qquad\qquad\qquad\qquad\qquad\qquad = \$4500$$
$$c = (1)(400) + (0.2)(750) + 635 \qquad\qquad\quad = 1185$$

operator/supervisor keypunching of rejects (10%)

$$\$5685$$

$$F = \frac{12,800,000}{5685}$$

$$= 2251 \text{ characters/dollar}$$

There are four major criteria for evaluating optical character readers. Cost, as discussed above, is the most obvious one, but it must be carefully related to the functional capabilities of data throughput speed, flexibility, and reliability. The higher speeds of buffered encoding devices and systems like the Mohawk data recorders must also be considered in a cost-per-dollar analysis. Naturally, all three of these capabilities directly influence the cost of character-reading equipment; but, as is the case with all equipment, the initial cost is only part of the story.

Throughput speeds are a function of reading speed, document transport speed, data density on the document, and multiline or page-reading capabilities. The rated reading speeds of current optical character readers range from about 200 to 2400 characters per second. When machines of different speeds and prices are compared, the number of characters read per dollar tends to increase at a much faster rate than machine costs.

Better performance in terms of flexibility and reliability might also save money over the long run, despite the higher initial equipment cost incurred. Flexibility pertains to a reader's ability to read a variety of character fonts, as well as its rescan ability (i.e., ability to reread a line of characters), paper-handling capability, and special format features. The ability to read only selected fields and to skip over crossed-out characters are two format features that are useful in some applications.

It may be that large quantities of in-house documents are created, retyped, or keypunched for the sole purpose of computer entry at a central station using a relatively small number of machines. If this is the case, the use of source-data recording equipment (such as Friden Flexowriters, Mohawk data-recorders, etc.) should be investigated as a possible alternative to OCR systems using RETYPE mode. The source-data recording devices produce computer-compatible paper or magnetic tapes, and thus eliminate the need for optical reading equipment. When the initial cost of installing the required number of source-data recording units approaches or exceeds the basic cost of an optical character reader, the latter method should be considered.

Special requirements of the application itself can be a determining factor. Certain legal considerations can support the use of OCR equipment in lieu of source-data recording devices. If a required hard-copy document is centrally produced, a case for OCR application exists. An alternative solution might be the use of typewriters that produce a paper-type or magnetic-tape record concurrently with the typing of the required documents. However, if the volume requirements are high enough, or if other company functions (billing, inventory, control, etc.) are amenable to OCR, it may be better to forego the use of paper tape or magnetic tape in favor of an OCR system, since paper tape is a relatively inflexible medium and magnetic tape cannot be deciphered by the unaided human eye.

Conversion to OCR of an in-house application, such as the simplified inventory control application shown in Figure 9-1, reduces the amount of direct human intervention in the data flow. This reduction can greatly increase data accuracy if there is a measure of quality control upon the production of the input documents, and if the documents are well designed for OCR processing techniques, as outlined in Chapter 4.

10. TRENDS AND FUTURE DEVELOPMENTS

The scope of applications for character readers is currently limited primarily by their inability to read a variety of different fonts, by their poor performance on handwritten documents, and by the lack of standardization within the industry. Consequently, there has been considerable development effort in these areas as well as in the areas of reliability and speed. The complexity of developing an OCR device has been reduced in the past few years by the commercial availability of packaged electronic systems—for example, minicomputers, shift registers, digital logic used for control functions, and peripheral equipment for keyboard entry and storage.

Experience in technologies related to OCR is providing several companies with a technological base for the development of an OCR product. Such technological propinquity has been introduced into several companies. For example, Information International has used its prior experience in manufacturing precision-scanning equipment for graphic arts composition to develop an OCR product. In addition, there have been about 40 research and development (R&D) projects, both commercial and government, in the past decade. There has been diffusion of the technology from projects at MIT, RAND Corporation, Sylvania Electronic Systems, Carnegie Institute of Technology, Stanford Research Institute, and RCA, among others. Moreover, many foreign post offices have spent large sums for R&D to develop recognition equipment for their mail-sorting problems. Such projects have produced technically trained people who have spawned or contributed to many commercial OCR enterprises.

114

An important result of these technological developments is the ease of entry into the ocr industry, illustrated by the numerous product announcements by new companies or existing companies without prior ocr experience.

COST REDUCTION IN ELECTRONIC COMPONENTS

Most of the modules that comprise an ocr system are almost entirely electronic. Therefore scanners, recognition units, and control units can be significantly reduced in cost as a result of integrated circuit technology. For example, one technique of scanning involves coupling an array of photoelectric sensors into an electronic storage register. One manufacturer recently provided this function with integrated circuitry at a cost of a few hundred dollars for 50 image elements. Just a few years ago, a discrete-circuit version of such a device would have cost well over $10,000.

The recognition unit of an ocr product typically consists of high-speed shift registers and special-purpose digital circuitry for comparison and control functions. Advances in electronic circuitry have significantly reduced the cost of this module too. For example, it is now possible to buy high-speed registers capable of storing 1000 image elements for approximately $400. A similar device would have cost $40,000 a few years ago. Further reduction in recognition cost can be achieved by replacing the special-purpose recognition logic with a less expensive, off-the-shelf minicomputer, and programming it for most of the recognition functions.

Minicomputers have also replaced special-purpose devices for the control units of ocr equipment. These small computers have been rapidly declining in price. Machines are now available at advertised prices of $3,000, and prices are expected to decline even further. Computers that cost well under $10,000 have processing capacity equivalents to that of devices costing as much as $100,000 a few years ago.

Another way to reduce cost of input/output is to use standard peripheral equipment available from independent manufacturers. For example, one ocr manufacturer charges $40,000 for a magnetic-tape transport and control unit, while available cassette storage systems can perform this function for under $1,000.

Another strategy—system compatibility with already installed equipment—was used in one document reader designed primarily for the bank-credit card industry. The reader has a printed output to facilitate other data-handling functions. It has been introduced by Data Recognition Corporation and costs approximately $80,000. The machine is used in conjunction with magnetic-ink character recognition (MICR) equipment. It optically reads account numbers from credit-card invoices and prints them in magnetic ink on the invoice at speeds up to 4000 cards per hour. The MICR equipment, which is already possessed by banks, sorts the cards by account number and thus eliminates the need for special OCR sorting equipment.

Lower Speed Yields Lower Transport Unit Cost

The document transport unit can represent as much as 50 percent of the total manufacturing cost of an OCR machine. There is a clear relationship between document-handling speed and system cost; lower-cost systems operate at lower speeds.

High speed, however, is not a requirement for a large percentage of users data preparation problems. This is indicated by the fact that current OCR machines, both page readers and document readers, are typically used only 2 to 4 hours per day. The maximum document-handling speed of the machines comprising the major share of installed OCR machines ranges from 7 to 20 documents per second. Allowing an average of 20 characters per document, the transport units of these machines are capable of processing the output of 70 to 200 keypunch operators. While high speeds are important in high-volume applications such as credit-card processing, the majority of all computer sites contain far fewer than 70 keypunch operators. Since only 6 percent of all computer sites contain over 70-keypunch operators, lower-speed machines can satisfy a large percentage of user problems.

Reduction in transport unit costs have been achieved in the newer products by reducing the speed of automatically fed systems and replacing automatic feeds with manual feeds. These approaches have been used by Datatype Corporation, Data Recognition Corporation, Infoton, Inc., and Orbital Systems, Inc.

A low-cost, hand-fed page reader introduced by Infoton in May 1970 operates at 12 lines per second, recognizes a stylized font, and sells for approximately $35,000. The system includes a cathode-ray tube display, which permits rejected characters to be displayed and corrected on-line by the operator.

Font Standardization vs. Multifont Capabilities

The "jack of all trades, master of none" theory was certainly applicable to recognition logic facilities in the past. Great sacrifices in reading reliability and increased cost resulted from the need to recognize a multitude of font styles. The first giant step toward standardization, however, was the acceptance of the ANSI standard character set for optical character recognition. Most of the subsequently designed OCR equipment contains facilities for reading this character set, and cooperation from users of business forms and manufacturers of data processing equipment has been very promising. The current trend toward lower-cost, single-font readers is bound to continue as forms design and special OCR ribbons, carbons, papers, and printers continue to improve. However, with the declining costs of integrated circuits used in the recognition unit, it will become less expensive to add multifont capability or to tailor special recognition logic for a new font. Though there has been an international effort to standardize fonts for optical character recognition, new products reflect a movement away from the proposed standards. It is becoming cheaper to introduce a flexible reading technology than it is to restrict the nature of the source documents characterizing the marketplace.

PROSPECTS FOR A NEW GENERATION OF OCR MACHINES

Many of the numerous OCR products introduced within the past year are dramatically different from earlier equipment. In a trend that is expected to continue, the new generation of OCR machines will be characterized by reduced cost and improved price/performance through use of integrated circuit technology, use of standard peripheral equipment available from independent manufacturers, employment of minicomputers for recognition logic and control functions, reduction in document transport speed, manual feed in lower-cost machines, and introduction of more special-purpose machines.

The number of off-line OCR systems that employ small general-purpose computers for software control of recognition and reformatting of data has definitely increased and will probably continue to do so. These satellite systems can place the Model 1401 commonly used with a larger 360 system to format input data before presentation to the large computer for processing. An OCR satellite system has available all the output possibilities of the controlling computer, and can communicate with the larger computer by compatible magnetic tape or any other high-speed storage medium, or by means of a communications or an on-line interface. Many of these systems are combination page-and-document readers capable of handling both documents and pages at respectable speeds. Three companies (Scan Data, CompuScan, and Information International) include omnifont capabilities.

A newer development receiving widespread interest is the remote transmission of OCR data (or ROCR) utilizing a facsimile transmission device that communicates with a recognition logic unit that generates magnetic tape. The Cognitronics Corporation employs a laser light source in the remote terminal, manually inserted corrections, and a time-sharing computer in its remote OCR system. The data converted by the recognition unit to magnetic tape is subsequently mailed back to the user or transmitted back over another data communication link to an incremental recorder on the user's premises. For recognition, the Cognitronics system uses a specially programmed PDP-8 computer with a CRT display of machine unrecognizable characters so that operators can enter these characters through a keyboard.

Recognition of Handwriting

Since each individual has his own style of handwriting, it is difficult to set recognition standards that will not lead to a high reject rate. Consequently this problem is even more perplexing than the multifont recognition problem because the recognition logic of the machine can never be set for a particular style.

The work being done on the recognition of handwritten characters may be divided into two classes: hand-printed characters and script. Some of the techniques being investigated in connection with handwritten documents are curve tracing, detection of selected features, and context recognition. Although a number of companies are working on the problem, most of the work has been kept confidential. The primary customer for a reader capable of handling handwritten documents appears to be the U.S. Post Office Department.

The IBM 1287 and 1288 optical readers can read both hand-printed numeric and alphabetic symbols, but a glimpse at the rigid set of rules for numerics (see Fig. 3-6) indicates that the concept is still quite restricted in scope. However, a number of other readers with hand-print recognition capabilities, including the Recognition Equipment Input II system and the Control Data Model 930-1, have recently appeared, attesting to the progress in this field.

APPENDIX I:
OCR AND PRINTER MANUFACTURERS

APPENDIX I:
OCR AND PRINTER MANUFACTURERS

OCR Manufacturers

Addressograph/Multigraph Corporation, 1200 Babbitt Road, Cleveland, Ohio 44117

Allied Computer Systems, Inc., 589 Boston Post Road, Madison, Conn. 06443

Burroughs Corporation, Second Avenue at Burroughs, Detroit, Mich. 48232

Cognitronics Corporation, 41 East 28th Street, New York, New York 10016

CompuScan, Inc., 125 Fort Lee Road, Leonia, New Jersey 07605

Control Data Corporation, Box 0, Minneapolis, Minnesota 55440

Cummins–Chicago Corporation, 4740 North Ravenswood Avenue, Chicago, Illinois 60640

Data Recogntion Corporation, 908 Industrial Avenue, Palo Alto, California 94303

Datatype Company, 1050 Northwest 163rd Drive, Miami, Florida 33169

Farrington Manufacturing Company, 6707 Electronics Drive, Springfield, Va. 22151

General Electric Company, P. O. Box 12313, Oklahoma City, Okla. 73112

Hewlett Packard, Cupertino Division, 11000 Wolfe Road, Cupertino, Calif. 95014

Honeywell Information Systems, Inc., Data Products Division, P. O. Box 2824, San Diego, Calif. 92112

Information International, Inc., 12435 West Olympic Boulevard, Los Angeles, Calif. 90064

Information Technology, Inc., 5070 Central Highway, Pennsauken, N. J. 08109

Infoton, Inc., Second Avenue, Burlington, Mass. 01803

International Business Machines (IBM), Data Processing Machines, 112 East Post Road, White Plains, N. Y. 10601

Motorola Instrumentation and Control, Inc., P. O. Box 5409, Phoenix, Ariz. 85010

National Cash Register Company, Dayton, Ohio 45409

National Computer Systems, 1015 South 6th Street, Minneapolis, Minn. 55415

OCR Systems, Inc., 328 Maple Avenue, Horsham, Pa. 94040

Optical Scanning Corporation, P. O. Box 40, Route 332 East, Newton, Pa. 18940

Orbital Systems, Inc., Church & Fellowship Roads, Moorestown, N. J. 08057

Philco-Ford Corporation, Communications & Electronics Division, Data Recognition Department, 3900 Welsh Road, Willow Grove, Pa. 19090

Recognition Equipment, Inc., 1500 West Mockingbird Lane, Dallas, Tex. 75235

Scan Data Corporation, 800 East Main Street, Norristown, Pa. 19401

Scan-Optics, Inc., 100 Prestige Park, East Hartford, Conn. 06108

UNIVAC Division, Sperry Rand Corporation, P. O. Box 8100, Philadelphia, Pa. 19104

PRINTER MANUFACTURERS

A. B. Dick Company, 5700 West Touhy Avenue, Chicago, Ill. 60648

Addmaster Corporation, 416 Junipero Serra Drive, San Gabriel, Calif. 91776

American Regitel Corporation, 870 Industrial Road, San Carlos, Calif. 94070

Bright Industries, Inc., One Maritime Plaza, San Francisco, Calif. 94111

Burroughs Corporation, Second Avenue at Burroughs, Detroit, Mich. 48232

Clevite Corporation, 37th and Perkins, Cleveland, Ohio 44114

Computer Devices, Inc., 167 Albany Street, Cambridge, Mass. 02139

Computer Measurements Co., 12970 Bradley Avenue, San Fernando, Calif. 91342

Connecticut Technical Corporation, 3000 Main Street, Hartford, Conn. 06126

Control Data Corporation, 8100 34th Avenue South, Minneapolis, Minn. 55420

Daconics, Inc., 505 West Olive Avenue, Sunnyvale, Calif. 94086

Data Computing, Inc. 2219 West Shangri-La Road, Phoenix, Ariz. 85029

Datamark, Inc., Cantiague Road, Westbury, N. Y. 11590

Data Printer Corporation, 225 Monsignor O'Brian Highway, Cambridge, Mass. 02141

Data Products, 6219 De Soto Avenue, Woodland Hill, Calif. 91364

Digital Equipment Corporation, 146 Main Street, Maynard, Mass. 01745

Dylaflow Business Machines Corp., 740 South Douglas Street, El Segundo, Calif. 90245

Eclectic Computer Corporation, 3707 Rawlins Street, Dallas, Tex. 75219

Electronic Information Systems, Inc., 2400 Industrial Lane, Bloomfield, Colo. 80020

General Electric Company, P. O. Box 12313, Oklahoma City, Okla. 73112

Gould, Inc., Graphics Division, 3631 Perkins Avenue, Cleveland, Ohio 44114

Gulton Industries, Inc., 15000 Central, East Albuquerque, N. Mex. 87108

Honeywell Information Systems, Inc., EDP Division, Peripheral Devices Operations, 300 Concord Road, Billerica, Mass. 01821

International Business Machines (IBM), Data Processing Machines, 112 East Post Road, White Plains, N. Y. 10601

International Computers, Ltd., OEM Sales and Technical Support, 839 Stewart Avenue, Garden City, N. Y. 11530

I/O Devices, Inc., 9 Skyline Drive, Montville, N. J. 07045

Automated Business Systems, Division of Litton Industries, 600 Washington Avenue, Carlstadt, N. J. 07072

Memorex Corporation, MRX Sales and Service Corporation Division, San Tomas at Central Expressway, Santa Clara, Calif. 95052

Mohawk Data Sciences Corporation, P. O. Box 630, Palisade Street, Herkimer, N. Y. 13350

Motorola Instrumentation and Control, Inc., P. O. Box 5409, Phoenix, Ariz. 85010

National Cash Register Company, Dayton, Ohio 45409

Nortec Computer Devices, Inc., Southboro, Mass. 01772

Novar Corporation, 2370 Charleston Road, Mountain View, Calif. 94040

ODEC Computer Systems, Inc., 871 Waterman Avenue, East Providence, R. I. 02914

Path Computer Equipment, Inc., 65 Commerce Road, Stamford, Conn. 06902

Potter Instrument Company, Inc., 151 Sunnyside Boulevard, Plainview, N. Y. 11803

Repco, Inc., 1940 Lockwood Way, Orlando, Fla. 32804

SMC Corporation, 299 Park Avenue, New York, N. Y. 10017

Shepard, Division of Vogue Instrument Corporation, 480 Morris Avenue, Summit, N. J. 07901

Syner-data, Inc., 133 Brimbal Avenue, Route 128, Beverly, Mass. 01915

Teletype Corporation, 5555 West Touhy Avenue, Skokie, Ill. 60076

Telex Computer Products, 6422 East 41st Street, Tulsa, Okla. 74135

Texas Instruments, Inc., P. O. Box 66027, Houston, Tex. 77006

UNIVAC Division, Sperry Rand Corporation, P. O. Box 8100, Philadelphia, Pa. 19104

Varian Data Machines, 2722 Michelson Drive, Irvine, Calif. 92664

Versatec, Inc., 10100 Bubb Road, Cupertino, Calif. 95014

APPENDIX II:
OCR AND PRINTER SUPPLIES
AND SUPPLIERS

APPENDIX II:
OCR AND PRINTER SUPPLIES
AND SUPPLIERS

BACKGROUND

When a computer printout is to be used for input to a character reader, a number of problems relating to type of paper and ink arise as a result of the methods of character recognition employed. The proper selection of supplies in OCR and MICR applications can determine whether or not the readers are economical, with acceptably low error-and-reject rates. For this reason, some suppliers have developed special OCR or MICR papers, inks, ribbons, and carbons.

Ink density in OCR printing may not be quite so critical as it is in MICR printing, but the reflectivity and color of the ink as well as the exactness of the character image are extremely important. Certain red or blue inks may be used to print NONREAD data, depending on the color of the light source. The difference between one blue ink that would be NONREAD and another that would be READ may not be discernible to the human eye. Furthermore, the relationship between the nonreflective element (ink) and the reflective element (paper) of character analysis may be blurred if the ink contains a reflective element (ink), and the reflective element (paper) of character analysis may be blurred if the ink contains reflective elements or if the paper does not reflect evenly. For the same reason, optical readers tend to be sensitive to dirt, document creases, and poor paper quality. Despite these drawbacks, OCR seems to offer the most promise for the future; and new techniques are being explored to overcome major functional problems.

127

SUPPLIERS

Supplies for ocr and micr printers consist mainly of ink, paper, carbons, pencils, and ribbons. Each of these is discussed separately in the subsequent paragraphs.

Ink

The ocr inks are of two kinds: READ and NONREAD. Both types are related to the light on the scanner of the individual reader, since the scanner light color will vary from company to company. NONREAD inks, which are generally a particular shade of red or blue, are those that have the same color as the scanner light and hence appear to the scanner as though they are the same color as the white paper. The READ inks, which are generally black, must be as non-reflective as possible, which is to say that they must contain as little of the pigmentation of the scanner-light color as possible. Since the critical differences between a READ and NONREAD red or blue or a "reflecting" and "nonreflecting" black are not discernible to the human eye, ocr inks should be analyzed and possibly be specially developed to fit the scanner.

Pencils

The development of special micr and ocr pencils relate to the same problems faced by ink manufacturers. Magnetic pencils and READ and NONREAD ocr pencils are used for correction on alterations of printed data and must generate marks that meet the same conditions as their corresponding inks.

Paper

Papers for ocr and micr installations must be heavy enough to be fed through a machine at rapid speed without jamming the transport as a result of curled edges, poor registration, tears, or double feeding. The ocr papers present a number of problems in addition to that of adequate handling weight. As indicated previously, the paper has to be tested for its reflectivity with respect to the particular scanner being used. Since character shape governs recognition, the ink must flow onto the paper smoothly, with a minimum of "feathering" at the edge of the character due to soaking by grains in the paper. Finally, nonwhite papers should be used with extreme caution because of the difference in contrast between the character and its background.

Carbons

Carbons used in ocr installations present a number of special problems in addition to the restrictions governing ink color and reflectivity. The sharpness of the character outline depends on the impact of the print hammers, the weight of both paper and carbons, and the number of carbon copies to be made. Manufacturers of carbon papers state that, given a suitable printing impact, the carbon

and paper weights have to be changed according to the number of copies desired. In other words, the carbon used to make a master and one copy would not
be suitable for making a master and four copies, and vice versa, if all are to be
used as input to an OCR reader. With suitable adjustments, however, carbon
copies can be made perfectly acceptable as OCR input. The use of copies produced by carbonless papers has had poor results with most OCR devices.

Ribbons

The stipulations governing ink also govern OCR and MICR ribbons. The OCR ribbons, since they must always generate a clear character, have a shorter life span
than normal printer ribbons of approximately the same quality. In general, it
saves money to use high-quality OCR ribbons.

FORMS

Many factors affecting costs and operational efficiency must be considered in designing and procuring the forms to be used in preparing OCR input. During the
initial system-design period and in subsequent changes, significant savings in
costs and improvements in efficiency can be achieved by keeping four basic
principles in mind:

1. Eliminate costly custom-form designs where practical by taking full advantage of the wide variety of standard forms and features available from the
major suppliers.
2. Avoid inefficient form layouts that result in slow effective printer speeds.
3. Standardize forms and schedule data processing runs to minimize the need
for frequent reloading of printers and adjustments for different forms sizes.
4. Eliminate the production of unnecessary documents, and increase the utility of the necessary documents by tailoring their formats and contents to the
needs of actual users.

Three distinct types of printers are used in computer installations. The primary computer printers are, of course, the high-speed line printers. Many computer systems also include typewriters for limited output such as logging of information, instructions to the operator, and console inquiries and responses.
Adding machines or listers are sometimes used (usually off-line) to provide list-
ings and control to totals. Tab printers generally use forms similar to those used
with the high-speed computer printers. Any of these printers can be used as OCR
devices where the printed copy is used as input to an optical character reader.

Continuous Forms

Continuous forms are used with nearly all computer and tab printers to permit
high-speed output and long runs without the need for continual operator atten-

tion, the most common type being pin-fed, continuous, fanfold forms. Pin-fed forms have sprocket holes punched into one or both margins to facilitate positive feeding and accurate registration.

Most typewriters use friction-fed forms without sprocket holes. Positive registration cannot be assured with this type of feeding mechanism, particularly when multipart forms are used. However, most typewriters can be adapted to handle pin-fed forms, through the use of either a modified carriage or an external device.

Multipart Forms

Where multiple copies of printed output are required, multipart forms may be tailored to meet a wide variety of needs. Two important considerations concerning the selection and use of multipart forms are the techniques used to transfer the printed images to the extra copies and to hold the copies together.

To transfer the images, separate sheets of carbon paper may be interleaved between the forms, or the back of each form may be coated with a carbon-transfer substance. The newest technique is the carbonless form, which requires a special paper that is treated to produce an image wherever pressure is applied. The force of the print hammers causes the printed characters to appear on each copy of the form.

Each image-transfer technique has its own advantages and disadvantages with regard to cost and ease of use. For example, interleaved-carbon forms are the most widely used and usually the cheapest, but additional effort is required to remove and dispose of the interleaved carbon paper when the forms are separated. Carbon-coated forms allow selected data to be deleted from particular copies by not providing a carbon backing in those areas. Carbonless forms are generally the easiest and neatest to use, but also the most expensive.

The type of fastening can also affect the ease of use and cost of forms. Three common methods are stapling, gluing, and crimping. The technique to be used depends on the characteristics of the printer and the facilities available for separating the forms. In general, the printer manufacturer's recommendations should be followed, to minimize the chances of damaging the printer or tearing the forms.

Special Forms

Many special types of forms are available to speed specific tasks. Examples of these forms include labels, envelopes, preprinted checks, and preprinted government forms such as Form W-2.

Some forms contain preprinted consecutive identification numbers. When buying such forms, it is wise to check whether the supplier will guarantee that the number sequence will contain no duplicated or skipped numbers. Errors of this type could seriously disrupt your reconciliation or forms accounting procedures.

In some applications, the number of legible copies that can be produced directly by the printer is insufficient, so the printed forms will need to be duplicated. Two specialized types of printer forms are available for specific use with off-line duplicating processes. One type is essentially the same as regular forms but uses a higher-quality paper to provide better resolution. This type is used where a separate photographic process is employed to produce plates, which are in turn used to print the copies. Advantages of the photographic processes are that reductions or enlargements in size can be made easily, high-quality printing is possible, and very large quantities of copies can be made. The second type is called a direct image master. The form printed on the computer serves as the master for the final duplicating process without any intermediate steps.

Continuous Roll Forms

Continuous roll forms or rolls of blank paper are frequently used for typewriters, adding machines, and listers.

The document-processing system developed by IBM permits continuous roll forms to be used by high-speed computer printers. Paper from a large roll, either blank or preprinted, is fed into the printer. After the documents have been printed, the paper is fed into a special device that cuts and stacks the forms. The feed roll contains over 8000 feet of paper—up to three times as many forms as are contained in a box of folded forms. This system does, however, require a significant amount of additional floor space at the printer.

Suppliers

The full address of each supplier is given in the following list.

Accurate Business Forms, Inc., 211 West Kilbourne Avenue, Milwaukee, Wis. 53203

Acme Visible Records, Inc., 126 W. Allview Drive, Crozet, Va. 22932

Addressograph-Multigraph Corporation, Marketing Research & Product Planning, 1200 Babbitt Road, Cleveland, Ohio 44117

Autopoint Company, Division of Cory Corporation, 3200 W. Peterson Avenue, Chicago, Ill. 60645

A. W. Faber-Castell Pencil Company, Inc., 41 Dickerson Street, Newark, N.J. 07103

Baltimore Business Forms, Inc., 3120 Frederick Avenue, Baltimore, Md. 21229

Bedinghaus Business Forms Company, Lippelman Road, Cincinnati, Ohio 45246

Buckeye Ribbon & Carbon Company, 7209 S. Clair Avenue, Cleveland, Ohio 44103

Burroughs Corporation, Lear-Siegler, Inc., 3000 N. Burcliek Street, Kalamazoo, Mich. 49003

Columbia-Great Lakes Corporation, P.O. Box 690, 1384 Highland Avenue, Cleveland, Ohio 44107

Columbus Ribbon & Carbon Manufacturing Company, Inc., Herb Hill Road, Glen Cove, N.Y. 11542

Continental Dataforms & Supply Company, 3812 N. Kedzie Avenue, Chicago, Ill. 60618

Control Data Corporation, Business Products Group, 4570 W. 77th Street, Edina, Minn. 55435

Copying Products, Division of Clopay Corporation, 417 E. 7th Street, Cincinnati, Ohio 45202

Data Products Corporation, 6219 Desoto Avenue, Woodland Hills, Calif. 91364

Data Speed, Inc., 274 Madison Avenue, New York, N.Y. 10016

Duplex Products Inc., 228 West Page Street, Sycamore, Ill. 60178

Dura Division, Intercontinental Systems, Inc., 2600 El Camino Real, Palo Alto, Calif. 94306

Eastman Kodak Company, Department 240, 343 State Street, Rochester, N.Y. 14650

Ennis Business Forms, Inc., 214 W. Knox Street, Ennis, Tex. 75119

Frye Manufacturing Company, P.O. Box 854, Des Moines, Iowa 50304

Graphic Controls Corporation, 189 Van Rensselaer Street, Buffalo, N.Y. 14210

IBM, Information Records Marketing, P.O. Box 10, Princeton, N.J. 08540

Joseph Dixon Crucible Company, 167 Wayne Avenue, Jersey City, N.J. 07303

Kores Manufacturing Corporation, 701 Whittier Street, Bronx, N.Y. 10474

The Label House, 155 W. Wisconsin Avenue, Pewaukee, Wis. 53072

Lewis Business Forms, Inc., 243 Lane Avenue North, Jacksonville, Fla. 32203

Moore Business Forms, Inc., Product Information Department, Suite 110-112 Parkway Inn Building, 401 Buffalo Avenue, Niagara Falls, N.Y. 14302

Olivetti-Underwood Corporation, 1 Park Avenue, New York, N.Y. 10016

Phillip Hano Company, 85 Sargeant Street, Holyoke, Mass. 01040

Pioneer Business Forms, P. O. Box 1237, Tacoma, Wash. 98401

RCA Corporation Computer Systems Division, Supplies Marketing, Cherry Hill, N. J. 08034

Reynolds & Reynolds Company, 800 Germantown Street, Dayton, Ohio 45401

Richard Best Pencil Company, 211 Mountain Avenue, Springfield, N.J. 07067

Safeguard Business Forms, De Marco Plant, Philadelphia, Pa. 19132

Singer Company, Friden Division, 2350 Washington Avenue, San Leandro, Calif. 94577

Standard Register, Dayton, Ohio 45401

UARCO Incorporated, W. County Line Road, Barrington, Ill. 60010.

Wallace Business Forms, Inc., 444 W. Grand Avenue, Chicago, Ill. 60610

Woehrmyer Printing Company, 301 York Street, Denver, Colo. 80205

GLOSSARY

Each new concept brings with it a specialized vocabulary of descriptive terminology. The most commonly used terms associated with optical character recognition are defined in the following alphabetical glossary.

Adjacency. A condition in which the character-spacing reference lines of two consecutively printed characters on the same line are separated by less than a specified distance.

Alphanumeric. Pertaining to a character set that includes both alphabetic character (letters) and numeric characters (digits). *Note:* Most alphanumeric character sets also contain special characters, such as punctuation or control characters.

Area Correlation. A technique of recognition in which the character space is divided into abstract cells (usually a rectilinear grid), and the presence or absence of ink in each cell is used to recognize the character.

AUTO-LOAD. A system (developed by Philco Corp.) that utilizes preprinted forms containing the scanning format and data control instructions. The AUTO-LOAD instructions, which function as the source program instructions, are read by the optical character reader into magnetic core memory immediately before reading a group of related source documents.

Background Reflectance. A measure of the amount of light reflected from the surface of a document. The amount of light actually reflected is compared with the amount of light that could be theoretically reflected from an absolutely white surface, and the result is expressed as a percentage of this absolute reflection. Generally, the lighter the color, the greater the reflectance.

Basic Weight. A standard measure of paper thickness and/or composition in pounds per ream (500 sheets) of 17x22-inch sheets (for bond), and 24x36-inch sheets (for card stock). Generally, the weights of OCR documents are 17 to 22 pounds for paper documents and 100 pounds for tabulating card stock.

Buffer. A device for the temporary storage of a single character or a fixed number of characters of data. The need for buffering generally arises from speed differentials between interfacing pieces of EDP equipment.

Carriage. The portion of a printing device that serves to hold and transport the paper being printed upon.

Cell Area. The area on the document being scanned at the instant of reading any given character.

Chain Printer. A line printer in which the type slugs are mounted on a chain that moves horizontally past the printing positions. *Note:* Chain printers generally provide more accurate vertical registration than the more commonly used drum printers, and interchangeable chains often permit rapid changes in the size or makeup of the character set.

Character. In optical character recognition, a single, uniquely shaped symbol used to represent data. Alphabetic characters are A through Z, numeric characters are 0 through 9, and control characters vary widely between character sets. *See also* Code.

Character Alignment. The placement of a printed or typed character in relation to its intended position. On most OCR devices, the horizontal or vertical character alignment should not vary by more than 50 percent of the character width or height, measured from the intended character position.

Character Boundary. The largest rectangle, with a side parallel to the document reference edge, each of whose sides is tangential to a given character outline.

Character Recognition. The indentification of graphic, phonic, or other characters by automatic means.

Character Set. A family of related character symbols. Character sets may be alphabetic, numeric, or both. Alphanumeric sets, sometimes called alphameric sets, are character sets that include alphebtic, numeric, and (usually) special or control characters.

Character Spacing. The center-to-center distance between adjacent characters. The two values often specified are the minimum and maximum distances permissible for proper recognition. Currently available OCR devices can accommodate spacings as small as 0.83 inch.

Character Spacing Reference Line. A vertical line used to evaluate the horizontal spacing of characters. It may be either a line that equally divides the distance between the sides of a character boundary or one that coincides with the centerline of a vertical stroke.

Clear Area. A specified area that is to·be kept free of printing or any other markings not related to machine reading.

Code. In optical character recognition, a complex of one or more printed symbols that represent a data unit by relative position rather than shape. *See also* Mark Sense.

Color. The spectral appearance of the image, dependent upon the spectral reflectance of the image, the spectral response of the observer, and the spectral composition of incident light.

Contrast. The difference between the color or shading of the printed material on a document and the background on which it is printed.

Control Character. A character whose occurrence in a particular context initiates, modifies, or stops an operation; e.g., a character that indicates the start of a field.

Document. A medium and the data recorded on it for human use; e.g., a report sheet or a book. Any record that has permanence and that can be read by a man or a machine.

Document Reference Edge. A specified document edge with which the alignment of characters is defined.

Dropout Colors. Specific colors of ink used for lines and instructions on OCR-designed forms. The purpose of such colors is to produce images that are invisible to certain optical scanners.

Drum. With reference to printing, the imprinting device is an on-the-fly printer, consisting of a constantly revolving shaft, drum, or series of interlocked wheels embossed with the characters that are to be imprinted.

Edit. To rearrange information. Editing may involve the deletion of unwanted data; the selection of pertinent data; the insertion of various symbols, such as page number and typewriter characters; and the application of standard processes such as zero suppression.

Embossment. The depth of print impression on a document. Most OCR devices permit embossment of up to 0.005 inch.

Error. In character recognition, the substitution of one character for another by the recognition logic. *See also* Reject.

Error Rate. The ratio of the number of documents in which one or more character substitutions occur to the total number of documents read. Most OCR devices maintain error rates on the order of 1 percent. *See also* Reject Rate.

Extraneous Ink. Any spot appearing within the cell area but outside the character itself, and caused by streaking, smear, or splatter.

Field. A specified area used for a particular category of data, e.g., a data entry block on a form containing a wage rate.

Field Selection. A control facility that permits selection of particular characters or character groups from the total data printed on the line being read.

Flying-Spot Scanner. A device employing a moving spot of light to scan a sample space. The intensity of the transmitted or reflected light is sensed by a photoelectric transducer.

Font. A family of graphic character representations (i.e., a character set) of a particular size and style.

Formatting. The ability of some optical scanners to rearrange data prior to output.

Frame. The total area of a single print position.

Grid. Two mutually orthogonal sets of parallel lines used for specifying or measuring character images.

Hard Copy. A visible record on a permanent medium.

Image Dissector. A mechanical or electronic transducer that sequentially detects the level of light in different areas of a completely illuminated sample space.

Intended Line. The line on the document on which the centers of characters should occur. It is normally parallel to on "aligning" edge of the document itself.

Invalid Character. A character that is unrecognizable by any given OCR device that is operating properly.

Light Stability. An image's resistance to color change when exposed to radiant energy.

Line Printer. A printer that prints all the characters comprising one line during each cycle of its action.

Line Skew. The condition in which a printed line is not parallel to the intended line.

Machine Cycle. The total time required to execute one or more parts of a total operation; e.g., transporting and reading two documents per second defines a 0.5-second machine cycle.

Magnetic Ink. An ink that contains particles of a magnetic substance whose presence can be detected by magnetic sensors. *See* MICR.

Mark Sense. Sensing of penciled marks on a page and transmission of the appropriate code (depending on the position of the mark on the page) to the recognition unit.

Mean Character Shape. The character shape whose electrical representation output from the recognition circuits exactly matches the stored representation that the machine uses in the match analysis for that character.

MICR. The machine identification of characters printed with magnetic ink.

MOCR. Microfilm Optical Character Recognition.

Mosaic. A defined geometric array in which character video information is stored prior to analysis by the recognition logic.

Multifont. Pertaining to the capability to recognize characters printed in more than one type font on a single page.

Noise. Any stray signal caused by either an undefined mark on the document or electrical interference.

Off-line. Operating independently of a computer.

Omnifont. Pertaining to the capability to recognize characters printed in any type font.

"One" Code. A binary pattern printed by a computer using the "1" key to indicate code positions.

On-line. Operating under the direct control and timing of a computer.

On-the-Fly Printer. A printer in which the type remains in motion during the printing process; at the appropriate instants during its movements, the paper and type are forced together to cause the desired character to be printed.

Opacity. The measure of a paper's ability to resist the passage of light.

Optical Character Recognition (OCR). The machine identification of printed characters through use of light-sensitive devices.

Optical Code Recognition. The machine identification of printed codes through the use of light-sensitive devices.

Paper Characteristics. The thickness, stiffness, tear strength, and other qualities of a paper stock that determine its suitability for use with any given OCR device.

Paper Grain. The direction of the fibers in paper composition. "Long grain" means that the long dimension of the document is parallel to the direction of paper stock movement through a paper-making mill. "Short grain" denotes that the short dimension of the document is parallel to the direction of paper travel through the mill.

Paper Smoothness. An arbitrarily established measure of a document's surface resistance to air flow. As the resistance decreases, the smoothness number increases. Low paper smoothness is a desirable quality in OCR applications.

Pitch. The horizontal distance between corresponding points of adjacent type characters; e.g., 12-pitch (12 characters per inch) is "elite" pitch, 10-pitch is "pica" pitch, and 8-pitch is "billing" pitch.

Platen. An element of the carriage in a typing or printing device, usually (but not necessarily) a hard rubber cylinder. The function of the platen is to support the paper as it is struck by the type face and to guide the paper as it is spaced.

Print Position. A position in which any one of the members of the printer character set can be printed in each line. *Note:* Most of the current line printers have between 80 and 160 print positions; i.e., they can print between 80 and 160 characters per line.

Record Format. The prescribed arrangement of an item or items of data to form a unit of related facts.

Registration. The physical positioning of a print line or character (vertical or horizontal registration) with relation to a form set or the machine itself.

Reject. To classify a document as unreadable, usually after several attempts have been made to read an unrecognizable character. The rejected document is generally routed to a reject bin or stacker. In some cases a visual display is made of the character for human interpretation, and/or a visual or audible alarm is activated. *See also* Error.

Reject Rate. The ratio of rejected documents to the total number of documents read.

Retina. An array of photoelectric transducers arranged in a grid to sense an entire character image simultaneously.

Scanning Format. Preprogrammed identification of fields to be read by an OCR device.

Skip or Slew. To move paper in a printer, without printing, through a distance greater than the normal line spacing, usually at a higher speed than in a single-line advance.

Solenoid. An electromechanical actuator used to convert electric energy into physical movement. In printers, solenoids are used to open the circuits that fire the print hammers.

Special Character. A character that is neither a letter nor a digit; it may be a punctuation mark (e.g., comma) or a control character that causes a particular operation to be performed (e.g., carriage).

Spectral Response. A device variation in sensitivity to light of different wavelengths.

Stroke. A straight line or curve between two nodes of a character. Each character is made up of a variable number of these strokes. Their shape and arrangement defines each individual character of a set. For a character to be read correctly, the quality of printing of each stroke must fall within certain tolerance limits, which are usually expressed in terms of stroke width.

Stroke Analysis. A technique of recognition in which the strokes and lines of a character are considered as descriptions of the character, as opposed to techniques in which the character space is divided into abstractly determined cells.

Stroke Average Width. The average of actual stroke widths taken at points along the length of a stroke.

Stroke-Edge Average. An imaginary smooth line that idealizes the true edge of a printed stroke.

Stroke Width. The nominal width of a stroke, usually expressed in decimal fractions of an inch (e.g., 0.012 inch). "Maximum stroke width" and "minimum stroke width" express, respectively, the maximum and minimum widths a stroke may have without adversely affecting the character recognition process.

Tractor. A device used on printers to control the vertical movement of paper through the carriage, normally by means of pinion wheels that engage pin-fed or punched-hole margins.

Vernier. A printer control, normally rotational in nature, used for fine vertical or horizontal carriage adjustments to align the form being printed while the printer is operating.

Vertical Format Control Tape. A punched paper or plastic tape, usually 8- or 12-channel, formed into a loop and used to control the spacing and skipping of a line-printer carriage.

Void. The inadvertent absence of ink within a character outline.

INDEX

143